9

Band 20 · September 1984

isotopenpraxis

AKADEMIE-VERLAG

ISSN 0021–1915 Isotopenpraxis, Berlin 20 (1984) 9, 321–364

EVP 9,– M

An unsere Autoren

Die Zeitschrift veröffentlicht Beiträge zu folgenden Themen: Kontrolle und Analyse radioaktiver und stabiler Isotope, Anwendung radioaktiver und stabiler Isotope in allen Gebieten der Forschung und Technik, der Industrie, Landwirtschaft und Medizin, Herstellung radioaktiver und stabiler Isotope sowie von markierten Verbindungen, kernelektronische Geräte für die Isotopenanwendung (Dickenmesser, Dichtemesser u. ä.) Detektoren für isotopentechnische Arbeiten und den Strahlenschutz in Isotopenlaboratorien, Meßtechnik und Strahlenschutz, Dosimetrie in Isotopenlaboratorien, kerntechnische Geräte für die Isotopenanwendung (Beton, Bestrahlungsanlagen u. ä.), Strahlenchemie, Radiochemie, ökonomische und perspektivische Fragen, Fragen der Weiterbildung.

1. Aufnahme von Manuskripten

Zur Veröffentlichung angenommen werden nur druckreife, noch nicht anderweitig publizierte Beiträge in deutscher, englischer oder russischer Sprache zum Themenbereich der Zeitschrift und zwar

Originalarbeiten
Kurze Mitteilungen
Aus der Praxis
Zusammenfassende Berichte
Tagungsberichte
Informationen

Sind der Beitrag oder Teile daraus bereits veröffentlicht oder ist eine anderweitige Veröffentlichung vorgesehen, so muß dies der Redaktion mit genauer Angabe der Zweitveröffentlichung mitgeteilt werden. Mit der Annahme des Manuskriptes behält sich der Verlag das Recht der Vervielfältigung, Verbreitung und Übersetzung innerhalb der gesetzlichen Schutzfrist vor. Auszugsweiser Nachdruck oder photomechanische u. a. Wiedergabe sind an eine schriftliche Genehmigung des Verlages gebunden. Manuskripte, Zeichnungen und Photos von veröffentlichten Arbeiten werden nur auf den besonderen und bis zum Zeitpunkt der Veröffentlichung mitgeteilten Wunsch des Autors zurückgesandt.

Bei der unverlangten Einreichung von Manuskripten wird empfohlen, eine Befürwortung eines der Herausgeber oder Redaktionsausschußmitglieder beizufügen.

Manuskripte sind einzureichen bei der Redaktion, 1086 Berlin, Leipziger Str. 3–4.

2. Form der Manuskripte

Die Autoren werden gebeten:

2 Manuskripte, eines davon in Original-Maschinenschrift, einseitig und mit 2zeiligem Abstand einzureichen; gut leserliche Photokopien sind zulässig; auf dem 1. Blatt (unter Weglassung der Bezeichnung)

den Titel der Arbeit
Vorname-Initialen und Namen sämtlicher Autoren
gegebenenfalls die Institution, in der sie tätig sind
eine kurze Zusammenfassung der Arbeit von höchstens 15 Zeilen
Keywords in alphabetischer Reihenfolge und
als Fußnote die Anschrift der Institution anzugeben;

wenn möglich, auf gesonderten Blättern Namen, Titel der Arbeit und Zusammenfassung in den beiden anderen obengenannten Sprachen 2fach beizufügen;

Literaturhinweise (fortlaufend numeriert), Fußnoten, Tabellen, Abbildungen (in 2–3facher Veröffentlichungsgröße), Abbildungsunterschriften (getrennt von den Abbildungen mit Zeichenerklärungen, Quellenangaben, falls erforderlich, und gegebenenfalls mit einem Vermerk für besondere Wünsche bezüglich der Abbildungsgröße), private und Dienstanschrift der Verfasser mit akademischen Titeln und ausgeschriebenen Vornamen auf besonderen Blättern ans Ende des Manuskriptes zu stellen;

Wünsche für den Kolumnentitel (Seitenüberschrift) bekanntzugeben;

Absätze durch Einzüge von mindestens 3 Anschlägen kenntlich zu machen; die Maßeinheiten des SI-Systems („Systeme International") zu verwenden; Korrekturen im Manuskript unmittelbar an der zu korrigierenden Stelle anzubringen;

Formeln deutlich lesbar und möglichst groß einzutragen;

Farbabbildungen und nachträgliche Änderungen oder Ergänzungen in Abbildungen zu vermeiden;

hervorzuhebende Textstellen zu unterstreichen, und zwar halbfetten Satz mit einer geraden schwarzen Linie, kursiven Satz (Schrägschrift) mit einer blauen Wellenlinie (alle Eigennamen). Sperrungen gestrichelt;

besondere Buchstaben durch farbige Unterstreichung zu kennzeichnen, und zwar Formelzeichen in

Griechisch-Kursiv ($\alpha, \beta, \ldots$) rot,
Kursiv ($A, B, \ldots$) blau
halbfetter Kursiv ($\boldsymbol{A}, \boldsymbol{B}, \ldots$) 2 Linien blau (Vektoren),
halbfetter Grotesk (**A**, **B**, ...) 2 Linien violett (Tensoren);

bei Literaturangaben auf Vollständigkeit zu achten. In nachstehender Reihenfolge müssen angegeben sein:

a) bei *Zeitschriften*: Vorname-Initialen und Namen sämtlicher Autoren, Titel der Zeitschrift (nach „Periodica Chimica" oder möglichst wenig gekürzt), Bandnummer (unterstrichen), Jahreszahl (in Klammern), Heftnummer (nur bei Zeitschriften, die jedes Heft mit S. 1 beginnen) sowie die Seitenzahl, auf der die zitierte Arbeit beginnt.

b) Bei *Büchern*: Initialen der Vornamen und Name des Autors (Herausgeber werden durch den Zusatz „ed." gekennzeichnet), voll ausgeschriebener Titel des Buches in der Originalsprache, evtl. Bandnummer, voll ausgeschriebener Name des Verlages, Nummer der Auflage (außer der ersten), Erscheinungsort(e), Erscheinungsjahr (hier nicht in Klammern setzen!) und evtl. Kapitelnummer, Seitenzahl u. ä.

c) bei *Sammelbänden, Tagungsberichten*: Es soll zusätzlich der Titel des zitierten Beitrags angegeben werden. Die Seitenzahl muß unbedingt angegeben werden. Aus dem Zitat muß hervorgehen, ob es sich um einen Tagungssammelband oder um einen noch nicht veröffentlichten Vortrag handelt. Zwischen Tagungsort und Erscheinungsort des entsprechenden Tagungsberichts ist zu unterscheiden.

3. Korrekturen

Der Autor erhält zwei Umbruchabzüge, von denen einer möglichst schnell korrigiert an die Redaktion zurückgereicht werden soll. Auf Wunsch des Autors kann die Korrektur auch von der Redaktion gelesen werden.

Durch nachträgliche Manuskriptänderungen verursachte Korrekturkosten gehen, sofern sie 10% der Satzkosten übersteigen, zu Lasten des Autors.

Der Autor erhält 50 Sonderdrucke kostenlos. Weitere Sonderdrucke können bis zur Rückgabe der Korrekturabzüge auf Kosten des Autors bei der Redaktion bestellt werden.

Bezugsmöglichkeiten

Bestellungen sind zu richten

— **in der DDR** an den Postzeitungsvertrieb unter Angabe der Kundennummer des Bestellenden oder an den AKADEMIE-VERLAG, DDR-1086 Berlin, Leipziger Straße 3–4, PF-Nr. 1233;
— **im sozialistischen Ausland** an eine Buchhandlung für fremdsprachige Literatur oder an den zuständigen Postzeitungsvertrieb;
— **in der BRD und Berlin(West)** an eine Buchhandlung oder an die Auslieferungsstelle KUNST UND WISSEN, Erich Bieber OHG, Wilhelmstraße 4–6, D-7000 Stuttgart 1;
— **in den übrigen westeuropäischen Ländern** an eine Buchhandlung oder an die Auslieferungsstelle KUNST UND WISSEN, Erich Bieber GmbH, Dufourstraße 51, CH-8008 Zürich;
— **im übrigen Ausland** an den Internationalen Buch- und Zeitschriftenhandel; den Buchexport, Volkseigener Außenhandelsbetrieb der Deutschen Demokratischen Republik, DDR-7010 Leipzig, Postfach 160; oder an den AKADEMIE-VERLAG, DDR-1086 Berlin, Leipziger Straße 3–4.

Zeitschrift „Isotopenpraxis"

Herausgeber: Im Auftrag des Akademie-Verlages herausgegeben von Prof. Dr. Justus Mühlenpfordt
Verlag: Akademie-Verlag, DDR-1086 Berlin, Leipziger Straße 3–4, PF-Nr. 1233; Fernruf: 2236221 oder 2236229; Telex-Nr.: 114420;
Bank: Staatsbank der DDR, Berlin, Kto.-Nr.: 6836-26-20712
Verantwortlich für den Inhalt des Heftes: Prof. Dr. H. Hübner, Leipzig
Anschrift der Redaktion (Fernruf Frau H. Schier: 6351177) der Zeitschrift Isotopenpraxis: Akademie-Verlag, DDR-1086 Berlin, Leipziger Str. 3–4, PF-Nr. 1233
Veröffentlicht unter der Lizenznummer 1527 des Presseamtes beim Vorsitzenden des Ministerrates der Deutschen Demokratischen Republik
Gesamtherstellung: VEB Druckerei „Thomas Müntzer", DDR-5820 Bad Langensalza
Erscheinungsweise: Die Zeitschrift „Isotopenpraxis" erscheint monatlich. Die 12 Hefte eines Jahrganges bilden einen Band. Bezugspreis je Band 216,– M zuzüglich Versandspesen. Preis je Heft 18,–
Bestellnummer dieses Heftes: 1079/20/9

AN (EDV) 12831

Isotopenpraxis 20 (1984) 9, S. 321 bis 329

Nuclear-Physical Methods of Analysis in Lunar and Planetary Investigations*

Yu. A. Surkov

(V. I. Vernadsky Institute of Geochemistry and Analytical Chemistry, USSR Academy of Sciences, Moscow)[1]

The advances of space technology ushered in a new era in the development of man's ideas about the solar system. The flights of space vehicles to the Moon, Venus, Mars and other planets made it possible to obtain quite new information on the composition, structure and properties of the matter of the solar system's bodies. In these studies a prominent role is played by remote and contact nuclear-physical methods of analyzing extraterrestrial matter — gamma-spectrometry, X-ray-spectrometry, alpha-spectrometry and neutron spectrometry. At present practically any space vehicle bound for planets carries scientific instrumentation based on the nuclear-physical methods of analysis. In this review paper we shall consider only gamma-spectrometry and X-ray-spectrometry by means of which the main information on the composition, structure and properties of the material of the Moon and planets was obtained.

Die Flüge von Raumschiffen zum Mond, zur Venus, zum Mars und zu anderen Planeten ermöglichten die Erlangung vollkommen neuartiger Informationen über die Zusammensetzung, die Struktur und die Eigenschaften der Materie der Himmelskörper des Sonnensystems. Eine hervorragende Rolle spielen bei diesen Untersuchungen die im Fern- und im Direktmeßverfahren angewandten kernphysikalischen Methoden zur Analysierung der extraterrestrischen Materie — die Gamma-, die Röntgen-, die Alpha- und die Neutronen-Spektrometrie. Im vorliegenden Überblick werden die Gamma- und die Röntgen-Spektrometrie betrachtet.

Keywords

gamma spectroscopy; lunar materials; Mars planet; Moon; planets; Venus planet; X-ray spectroscopy

1. Gamma-spectrometry

In the early 1960s, when the first Soviet space vehicles for exploring the Moon were prepared, a question arose: how was it possible to determine the composition of the lunar crust? The possibilitiy of using various remote methods of analysis was considered, but finally gamma-ray-spectrometry was chosen. The Moon's first satellite — Luna-10 launched in 1966 carried the first gamma-spectrometer (Fig. 1) which determined from the gamma-radiation of lunar rocks the content of naturally occurring radioactive elements — uranium, thorium and potassium in them [1].

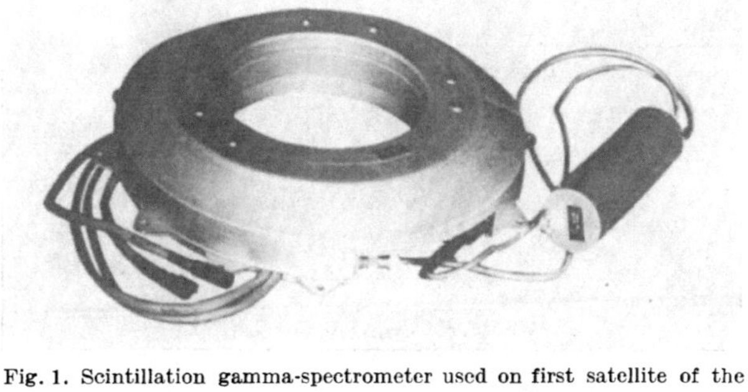

Fig. 1. Scintillation gamma-spectrometer used on first satellite of the Moon "Luna-10"

Proceedings from the content of radioelements the rocks occurring on the lunar space were classified and ideas about their origin and formation were developed. This experiment marked the beginning of the use of nuclear-physical methods of investigating extraterrestrial matter by space vehicles. Since then gamma-spectrometry has been repeatedly used for determining the composition and radioactivity of rocks on the Moon, Venus and Mars [2].

Tab. 1 lists spacecrafts which used gamma-spectrometers and areas where measurements were made. As evident from the table, in all cases scintillation spectrometers were applied. They have high efficiency and are simpler in design than spectrometers with semiconductor or gas detectors.

Tab. 2 lists the main characteristics of the gamma-spectrometers which were used in lunar and planetary studies.

a) *Radioactive elements on the Moon*

The greatest amount of information on gamma-spectrometry was obtained on the Moon. The natural radioactivity of lunar rock was measured in natural bedding from Luna-10 and Luna-12 and Apollo-15 and Apollo-16 orbital vehicles and in soil samples delivered from the Moon to Earth by Lunas 16, 20 and 24 and by American spacecraft Apollos 11, 12, 14, 15 16 and 17. The results of investigations indicate a rather large range of changes in the uranium, thorium and potassium content for various regions of the lunar surface or individual lunar soil samples brought to Earth: from 0.1 to 5 g/t for uranium, from 0.4 to 18 g/t for thorium and from 0.03 to 0.7% for potassium. (In some lunar soil samples the contents go beyond these limits.)

Fig. 2 presents the outline map of the lunar surface. It indicates the areas where the radioactivity of rocks was measured by Lunas 10 and 12 [2] and Apollos 15 and 16 [3] and also the regions in which the samples studied on the Earth were taken.

*) Plenary Lecture, presented at the 3rd Symp. on Nuclear Analytical Methods, April, 11–15, 1983 in Dresden

[1]) Address: 117975, GSP-1 Moskau W-334; Worbjewski Ch., Haus 47a

Tab. 1. Space vehicles and scientific instruments used to study the radioactivity of the Moon and planets

Launch year	Spacecraft	Area of measurements	Type of the instrument
1966	Luna-10	lat. ±72° long. 0 – 360°	32-channel scintillation spectrometer
1966	Luna-12	lat. ±20° long. 0 – 360°	the same
1971	Apollo-15	lat. ±29° long. 0 – 360°	direct transmission
1972	Apollo-16	lat. ±10° long. 0 – 360°	the same
1972	Venera-8	lat. – 10° long. 304°	64-channel scintillation spectrometer
1974	Mars-5	lat. – 50 + 20° long. 0 – 150°	256-channel scintillation spectrometer
1975	Venera-9	lat. 32° long. 291°	64-channel scintillation spectrometer
1975	Venera-10	lat. 16° long. 291°	the same

Tab. 2. The main characteristics of the gamma-spectrometers which were used for lunar and planetary studies

Spacecraft	Luna-10 Luna-12	Apollo-15 Apollo-16	Venera-8	Venera-9 Venera-10	Mars-5
Size of the NaI (Te), mm	40 × 40	76 × 76	63 × 63	63 × 63	63 × 63
Number of channels	32	–	64	4	256
Channel capacity, imp.	2^{12}	–	2^{15}	2^{12}	2^{16}
Resolution, %	10	8.5	10	10	10
Mass, kg	3.9	9.7	6	3.5	2.5
Power consumption, W	2.5	1.6 21.6 (with thermo-static control)	2.5	2.1	5
Range of energy measured, MeV	0.3 – 3.2	0.55 – 8.6	0.3 – 3.2	1.15 – 1.55 1.55 – 2.31 2.3 – 3.0	1.0 – 9.0

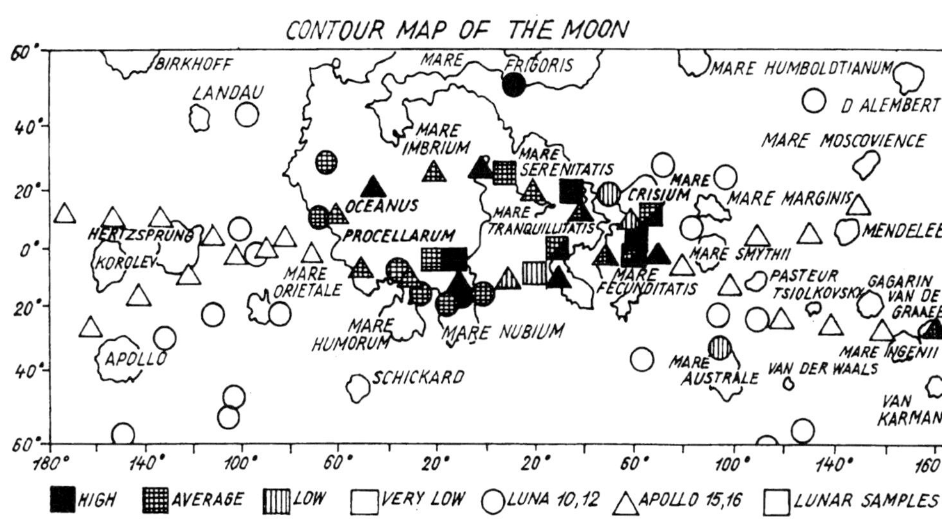

Fig. 2. Contour map of the Moon with the regions of measurements of the radioactivity

As evident from the map, the highest radioactivity takes place in the western basin of the lunar maria, radioactivity is lower on eastern maria and the lowest on continents and highlands of the Moon's visible and far sides. Such distribution of radioactivity became clear after studying lunar soil samples on the Earth. It is explained by the presence of two fundamentally different types of substance on the Moon's crust: the continental type (of the gabbroanorthosite type) and the mare type (of the basaltic composition). In highlands there is also the CREEP basaltic rock which has a high content of natural radioelements. However, it is little spread.

Tab. 3 lists the average contents of natural radioactive elements (uranium, thorium and potassium) for the main types of lunar rocks measured by gamma-spectrometry.

Tab. 3. The U, Th and K content in lunar rocks

Type of rock	U, 10^{-6}	Th, 10^{-6}	K, %
Basalts (CREEP)	2·8	9·3	0·57
Mare basalts (average)	0·64	2·61	0·15
Highland gabbro-anorthosite rocks	0·43	1·13	0·009
Range of contents in rocks of the lunar crust	0·1 — 5·0	0·4 — 18	0·03 — 0·7

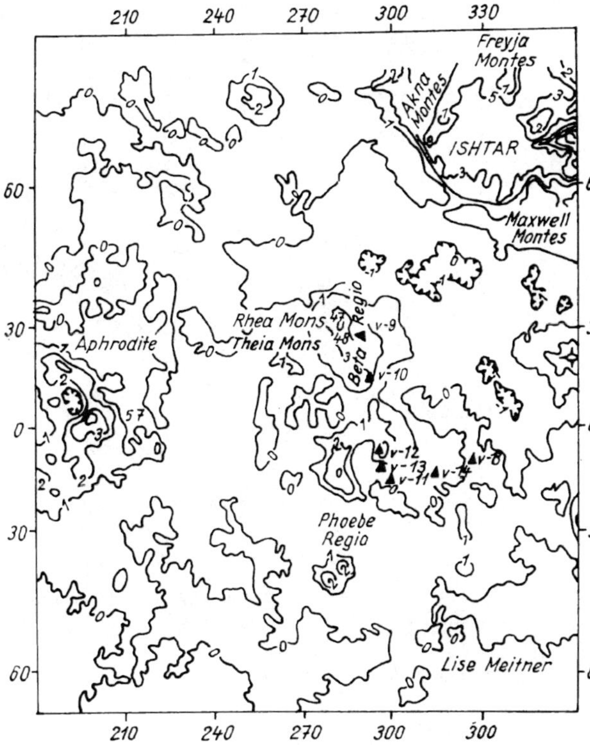

Fig. 3. Radar map of Venus taken by Pioneer-Venus spacecraft on which the landing sites of space probes "Venera" are shown

The low contents of radioelements on the Moon show that the processes of the differentiation of matter on the Moon were not so intensive as on the Earth and therefore the enrichment of the crust with radioelements on the Moon proved to be by an order of magnitude lower than on the Earth.

b) *Radioactive elements on Venus*

Unlike the Moon Venus has a dense atmosphere which rules out the possibility of studying the radioactivity of its rocks from orbiting vehicles. That is why the uranium, thorium and potassium content was measured by landing modules of Veneras 8, 9 and 10. In these experiments the Venusian atmosphere played a positive role since it absorbs cosmic rays and reduces the background of the gamma-radiation induced by them.

Fig. 3 presents the map of the Venusian surface based on radar results obtained by the altimeter installed on Pioneer-Venus [4]. The map indicates the landing sites of Soviet Venera probes on which nuclear-physics methods of analysis were used.

Tab. 4 lists the uranium, thorium and potassium contents measured by the gamma-spectrometer carried by Veneras 8, 9 and 10 [2]. As evident from Tab. 4 for various regions of the Venusian surface the uranium content varies by 4—5 times, while the thorium and potassium content varies by 10 times.

At the landing site of Venera-8 the rock has the highest content of natural radioelements which surpasses the average level of their contents in terrestrial basalts and is close to highly alkaline basalts. Perhaps this region of the Venusian surface where Venera-8 was landed is a section of the planet's ancient crust.

Data obtained by Venera-9 and Venera-10 make it possible to relate the investigated rocks to rocks of the transitional type between tholeitic basalts and alkaline basalts, that is, to rocks which are the product of the relatively not deep differentiation of the planet's primordial matter. These rocks seem to be younger volcanic formations than the rock at Venera-8 landing site. So, on Venus like on the Moon we see two types of geologomorphological provinces quite different in their nature or two types of the crust.

Tab. 4. The K, U and Th content in Venusian rocks

Probes and types of rocks	U, 10^{-6}	Th, 10^{-6}	K, %
Venera-8	2·2 ± 0·7	6·5 ± 0·2	4·0 ± 1·2
Venera-9	0·60 ± 0·16	3·65 ± 0·42	0·47 ± 0·08
Venera-10	0·46 ± 0·26	0·70 ± 0·34	0·30 ± 0·16

The magnitude of the K/U ratio for investigated Venusian rocks proved to be very close to the magnitude of the ratio typical of the Earth's magmatic rocks (about 10^4). It can be concluded from this fact that if processes took place on Venus leading to the change in the chemical composition of primordial material (no matter what was the nature of these processes), in the main they resembled terrestrial processes.

c) *Radioactive elements on Mars*

The first and so far the only measurements of Martian gamma-radiation were conducted from Mars-5 in 1974 [5] from the distance of about 1700 km from the planetary surface. That is why the measured spectra are averaged on the vast territory of the Martian surface. Fig. 4 shows this territory. (Gamma-radiation from the surface limited by the contour accounted for 90% of entire gamma-radiation recorded by the detector). The measurements were

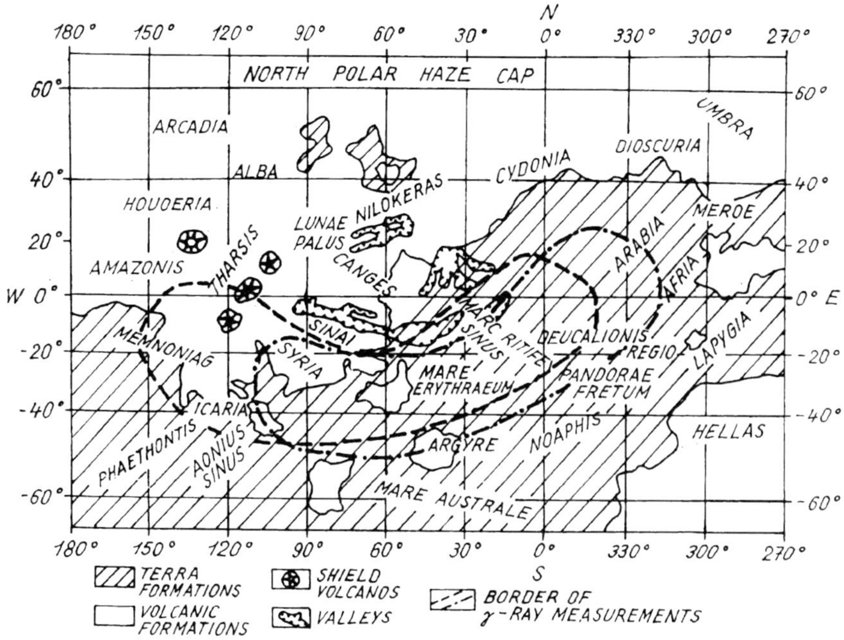

Fig. 4. Contour map of Mars with the regions of measurements of the radioactivity by "Mars-5" orbiter

made by the 256-channel scintillation gamma-spectrometer whose characteristics are given in Tab. 2.

As a result of processing measured gamma-spectra the content of natural radioactive elements was determined for the entire region explored and individually for two main types of geological and morphological provinces — ancient highland formations and younger volcanic areas. The data obtained are summarized in Tab. 5.

Tab. 5. The U, Th and K content in Martian rocks

Type of rock	U, 10^{-6}	Th, 10^{-6}	K, %
Planetary surface (averaged data)	0.6	2.1	0.3
Young volcanic rocks	1.1	5.0	—
Ancient highland rocks	0.2	0.7	—

The content of naturally occurring radioelements is a geochemical indicator of the type of rock and can be used for its identification. Therefore, noticeable differences between Martian volcanogenic and highland formations as far as the thorium and uranium content is concerned should apparently be interpreted as differences in the overall geochemical type of material forming them. At the same time the similarity of investigated Martian volcanites in the thorium and uranium content with basalts of some formations of the Earth and the Moon should be regarded in a wider respect — as the similarity of the types of rocks, which is in good agreement with the typically „basaltic" morphology of these formations. Finally, the similarity of the rocks of Martian highland provinces in the thorium and uranium content with gabbro-anorthosite rocks typical of lunar highlands well agrees with the fact that in their morphology and structure Martian crater-studded highlands are very similar to lunar highlands.

Summing up the consideration of the results of the gamma-spectrometric studies of the Moon and the planets, let us turn to the K-U diagram.

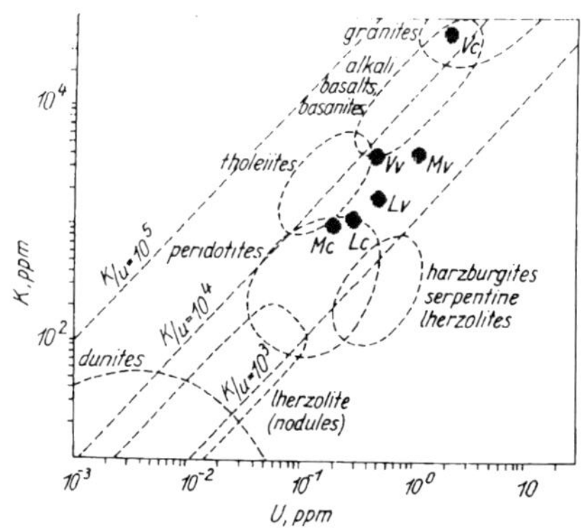

Fig. 5. K-U-diagram of the U, Th and K concentrations in the main types of the rocks of the Earth, Venus, Mars and the Moon

L_c — lunar continents; L_v — lunar maria; V_c — ancient crust on Venus; V_v — volcanic formation on Venus; M_c — ancient crust on Mars; M_v — volcanic formation on Mars

Fig. 5 gives the K-U diagram for the main types of the rocks of the Earth's crust [6] and also for the two main varieties of the crust of the Moon, Mars and Venus. It is evident from the figure that the K/U relations for the Moon and the planets are grouped in the vicinity of K/U relations typical of the most widespread tholeitic and alkaline basalt rocks of the Earth. On the Moon, as the least differentiated body, the difference between the average contents and the ratios of radioelements between ancient highlands and younger maria is the least. On Mars the difference is somewhat greater. On Venus the difference in the content and the ratio of radioelements is still greater and is close to that which takes place on the Earth.

Thus, it can be noted that one of the fundamental features of the Moon and the terrestrial planets is the presence of two main geomorphological types of the surface formed by two different types of matter. Ancient highland regions consist of feldspar rock: gabbroanorthosite on the Moon (and perhaps on Mars) and granite-metamorphic rock on the Earth (and perhaps on Venus). Younger "sea" areas consist of basalt rock. The formation of these two types of the crust took place in the course of different processes and during different periods of the evolution of the planets.

2. X-ray spectrometry

In its possibilities the X-ray spectrometric analysis of extraterrestrial material perhaps ranks second (after gamma-spectrometry) as a method used in planetary exploration. Using this method, it is possible to determine from the orbit of the satellite of the planet (if the latter is devoid of the atmosphere) and from the landing module the content of the basic rock-forming elements and rare elements in the surface rocks of planets and satellites. The first simplest experiment in studying lunar X-ray radiation was performed at the Moon's artificial satellite Luna-10 in 1966 by *Mandelshtam* et al. [7]. Somewhat later more complicated onboard instrumentation with (T-Ti) and (T-Zr) sources was used by *Kocharov* et al. [8] for an X-ray radiometric analysis of lunar soil on self-propelled "Moon cars" Lunokhod-1 (1970) and Lunokhod-2 (1973). In 1972 *Surkov* et al. [9] considered for the same purposes the possibility of analyzing extraterrestrial matter by a dispersionless X-ray fluorescent analyzer using a source of alpha-

Tab. 6. Spacecraft and scientific instruments used for analysis of the composition of lunar and planetary matter

Launch year	Spacecraft	Region of measurements	Type of the instrument
1966	Luna-10	lat. ±72° long. 0−360°	X-ray radiation detectors
1970	Lunokhod-1	Mare Imbrium N. L. 38°17′ W. L. 35°00′	64-channel X-ray spectrometer
1973	Lunokhod-2	LeMonnier crater N. L. 25°51′ E. L. 30°27′	64-channel X-ray spectrometer
1972	Apollo-15	lat. ±29° long. 0−360°	8-channel X-ray spectrometer
1972	Apollo-16	lat. ±10° long. 0−360°	8-channel X-ray spectrometer
1973	Viking-1	N. L. 22°48′ W. L. 47°82′	128-channel X-ray spectrometer
1973	Viking-2	N. L. 47°97′ W. L. 225°57′	128-channel X-ray spectrometer
1978	Venera-12	cloud layer (H = 47−63 km)	256-channel X-ray spectrometer
1981	Venera-13	lat. −7°30′ long. 303°11′	256-channel X-ray spectrometer
1981	Venera-14	lat. −13°15′ long. 310°09′	256-channel X-ray spectrometer

Tab. 7. The basic characteristics of X-ray spectormetric instrumentation used for lunar and planetary exploration

Spacecraft	Lunokhod-1 Lunokhod-2	Apollo-15 Apollo-16	Viking-1 Viking-2	Venera-12	Venera-13 Venera-14
Type of detector	proport. counters Ar + CH_4	proport. counters CO_2 + He	proport. counters Xe + Ne + He	proport. counters Kr + CO_2	proport. counters Kr + CO_2
Number of channels	64	8	128	256	256
Channel capacity (pulses)	—	—	2^5	2^{16}	2^{16}
Resolution, %	22−26 by $E_x = 4{\cdot}5$ keV		~25 by $E_x = 5{\cdot}9$ keV	20−25 by $E_x = 5{\cdot}9$ keV	20−25 by $E_x = 5{\cdot}9$ keV
Mass, kg				10	8
Power, consumption, w				34	9
Range of determined elements	from Al to Fe	Mg, Al	from Mg to Fe	from Al to Hg	from Mg to Fe
Excitation source of X-radiation	(T-Ti) or (T-Zr)	solar radiation	$^{55}Fe + ^{109}Cd$	$^{55}Fe + ^{109}Cd$	$^{55}Fe + ^{238}Pu$

particles of plutonium-238. In 1972 *Adler* et al. [10] described experiments in measurements of fluorescent radiation from the orbital modules of Apollos 15 and 16. In 1973 the description of instrumentation and the results of an X-ray fluorescent analysis of Martian rock by spacecraft Vikings 1 and 2 was given by *Clarke* et al. [11]. In 1978 and 1982 *Surkov* et al. [12, 13] described experiments in an X-ray fluorescent analysis of Venusian rock and aerosol of Venusian clouds by Veneras 12, 13 and 14.

Tab. 6 lists spacecraft and instruments which carried out X-ray fluorescent analysis of lunar and planetary material while Tab. 7 summarizes the basic characteristics of the instruments which have been used.

a) *X-ray fluorescent analysis of lunar rocks*

In this section two experiments in an X-ray fluorescent analysis of lunar rocks conducted by Lunokhods 1 and 2 and Apollos 15 and 16 will be considered. The results of these experiments are an important part of new scientific information about the Moon obtained in the 1970s.

Exploration by Lunokhod 1 and 2. The objective of Lunokhods 1 and 2 was to study the geological structure of the lunar surface, the character of rocks and the conditions of their formation. Lunokhod-1 conducted investigations at Mare Imbrium, i. e., in the region typical of most lunar maria, while Lunokhod-2 operated in the LeMonnier crater in the outlying region of Mare Serenitatis.

X-ray fluorescent spectrometers with radioisotopic sources were installed on the Lunokhods and determined the composition of lunar rocks [8]. The spectrometer consisted of two parts situated inside and outside the Lunokhod's

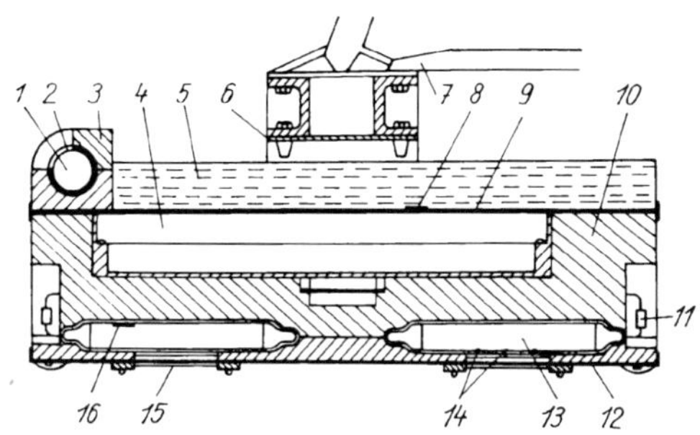

Fig. 6. Detection unit of the X-ray fluorescence spectrometer used on the "Lunochod-2" spacecraft

1 — X-ray detectors of the Sun; 4 — preamplifier; 5 — heat accumulator; 8 and 16 — temperature sensor; 13 — X-ray detectors of lunar soil; 2, 3, 6, 7, 9—12, 14, 15 — construction elements of the detection unit

pressurized compartment. The outlying electronic unit (Fig. 6) was located outside. It contained the radiation detector, preamplifiers, the commutator and panels with radioisotopic sources. The pressurized compartment housed electronic units which contained the pulse-height digital converter and the encoder.

The spectrometer had 10 detectors-proportional counters and two panels with radioisotopic sources. Lunokhod-1 used (T-Ti) sources on the molybdenum backing with the activity of 900 curies, while Lunokhod-2 made use of (T-ZR) sources on the tungsten backing with the activity of 4000 curies.

The detectors and the sources were placed in front of the Lunokhod about 30 cm from the soil surface. From this distance, if collimators were available, the detectors recorded fluorescent radiation from the surface section confined by the ellipse with 1·2 m and 0·8 m axes.

Tab. 8 summarizes the results of studying the composition of lunar rocks by Lunokhods 1 and 2 [8]. The results of analyzing rock along the route of Lunokhod-1 show that the variations in the elemental composition of rocks at different points are small and lie within the limits of the accuracy of measurements. Tab. 8 lists average data.

Tab. 8. The composition of lunar rocks measured by Lunokhods 1 and 2

Element	Lunokhod-1 (average composition)	Lunokhod-2	
		mare part of the route	highland part of the route
Al	7	8·8	11·6
Si	20	24	22
K	<1	<1	<1
Ca	8	8·0	9·1
Fe	12	6·1	4·1

Three regions can be singled out on the surface explored by Lunokhod-2: the lowland of the LeMonnier crater floor, the highlands and the tectonic fissure zone. The results of analyzing the rocks on the lowland part of the LeMonnier crater give grounds to believe that the rocks here are not typically mare ones since the iron content (6%) in them (see Tab. 8) is much lower than in typical mare regions where it reaches 10—12%. In the highlands in the upper layer of regolith the content of elements (including iron) proved to be close to that in other highland regions in the Moon. The results of the analysis show that in the LeMonnier crater area mare matter mixes with highland matter and there is a gradual transition from one type to the other with the motion of Lunokhod-2.

Investigations by Apollos 15 and 16. In 1971 and 1972 Apollos 15 and 16 were launched and from their orbital modules the lunar surface's fluorescent radiation excited by solar X-rays was measured. The measurements were conducted from the circular orbits of the Moon's satellite 100 to 120 km away from the lunar surface.

Solar X-ray-emission has the highest intensity in the 1 to 3 keV region. Therefore, it effectively excites fluorescent radiation of only Mg, Al and Si. To measure the fluorescent radiation of these elements X-ray spectrometers with a small number of channels were used. Three porportional gas counters with beryllium windows 25 microns thick and with the effective area of about 25 cm² served as detectors in the spectrometers. Two counters operated with filters (Mg 12·7 microns thick and Al 5 microns thick) and one counter had no filter. The angle of view of the detectors was limited by the collimator and was equal to 60°. After their amplification and shaping the signals from the detectors entered the eight-channel pulse-height analyzer the data from which were transmitted to Earth via telemetry channels.

The channel of the spectrometer without a filter operated in two modes: 0·75—2·75 keV and 1·5—5·5 keV. (The switching was automatic.) The channels of the spectrometer with detectors provided with filters operated in the 0·75 to 2·75 keV energy range. To control solar activity a special detector was used. It operated in the 1 to 3 keV range.

As a result of the experiment the relations of the intensity of fluorescent radiation (and hence the Mg/Si and Al/Si ratios) were measured. It was established that the ratios of these elements vary by 1·5 to 2·0 times with the transition from mare rocks to highland rocks. In mare regions the Al/Si ratio is lower and the Mg/Si ratio is higher than in highlands. These facts become clear, being compared with the results of studying the composition of lunar soil samples brought to Earth. They show that lunar maria consist of basaltic rocks which have a higher Al content and a lower Mg content the gabbro-anorthosite rocks of lunar highlands.

b) *X-ray fluorescent analysis of Venusian rocks*

Investigations by Veneras 13 and 14. Venera-13 and Venera-14 reached Venus in March 1982. The landing sites were chosen so that data on the topography and composition of rocks for the most typical provinces of the Venusian surface could be obtained. Venera-13 explored rock in the upland rolling plain, while Venera-14 studied rock in the lowland with the smooth relief. These two types of Venusian provinces account for more than 80% of the planetary surface. That is why the probes' data on the composition of rocks can be considered sufficiently representative for the Venusian surface. Fig. 7 shows the location of the spectrometer on the landing module, while Fig. 8 shows the geometry of measurements. The instrument consists of two units: the detection unit and the multichannel pulse analyzer. Both units are housed in the hermetically sealed thermostated compartment of the landing module. The Venusian rock sample for analysis was taken by a drilling device, was transported through air-lock device 5 and soil sampler 8 and irradiated by radioisotopic sources 9. The fluorescent radiation thus appearing is recorded by detectors 4 the signals from which enter through preamplifiers into multichannel pulse analyzer 1. Information accumulated in the analyzer is periodically sent to the telemetry system for transmission to Earth.

The device uses a radioistopic ^{238}Pu source with the activity of 50 mc and two ^{55}Fe sources with the activity of 125 mc each. The use of two types of radioisotopic sources ensures sufficiently high efficiency within the wide range of the elements to be determined (from Mg to Fe).

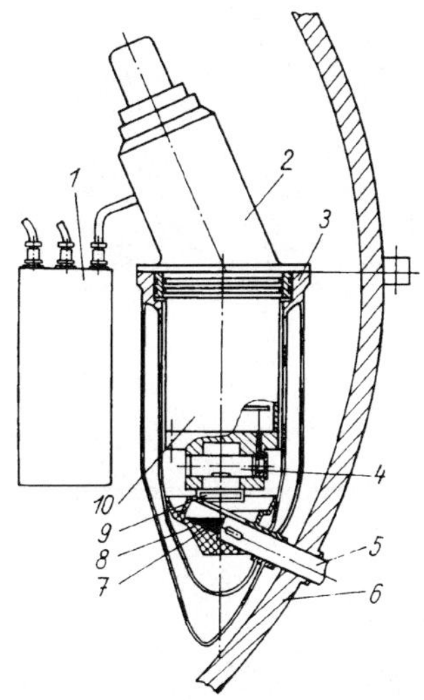

Fig. 7. Location of the X-ray fluorescence spectrometer on the landing module of Venera 13 and 14

1 — puls hight analyzer; 2 — lid of the detection unit; 3 — casing of the detection unit; 4 — counters; 5 — air lock; 6 — hull of the landing module; 7 — sample under study; 8 — rock receiver; 9 — radioisotopic sources; 10 — electronic section

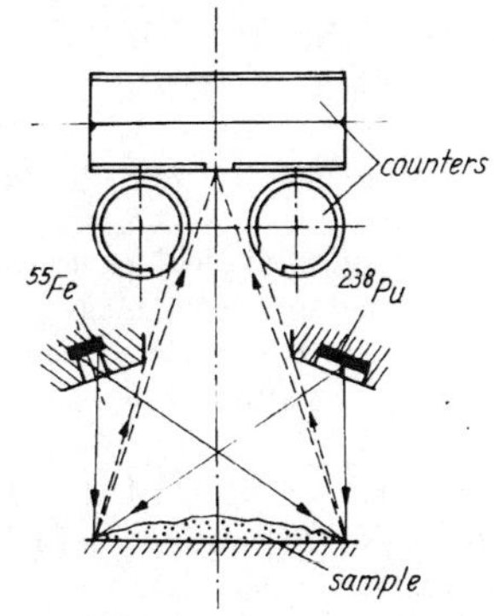

Fig. 8. Diagram showing relative location of sample under study, radioisotopic sources, and proportional counters

The fluorescent radiation of rock is recorded by our detectors — gas-discharge proportional counters. The mixture of 90% of Kr and 10% of CO_2 is the counters' working gas. The counters have beryllium windows 40 microns thick. The basic characteristics and parameters of the X-ray fluorescent spectrometer are given below:

The mass	8 kg
Power consumption	9 kg
The number of the analyzer's channels	256 (2 by 128)
The capacity of the analyzer's channel	$2^{16} - 1$
The range of energies measured	1·1 to 8 keV
Energy resolution	20 to 25% (along the line of 5·9 keV)
The range of the basic elements to be determined	from Mg to Fe

Tab. 9. The composition of rocks measured by Veneras 13 and 14 (in weight percent)

Elements (oxides)	Venera-13 (upland)	Venera-14 (lowland)
MgO	11·4 ± 6·2	8·1 ± 3·3
Al_2O_3	15·8 ± 3·0	17·9 ± 2·6
SiO_2	45·1 ± 3·0	48·7 ± 3·6
K_2O	4·0 ± 0·63	0·2 ± 0·07
CaO	7·1 ± 0·96	10·3 ± 1·2
TiO_2	1·59 ± 0·45	1·25 ± 0·41
MnO	0·2 ± 0·1	0·16 ± 0·08
FeO	9·3 ± 2·2	8·8 ± 1·8
SO_3	1·62 ± 1·0	0·88 ± 0·77
Cl	<0·3	<0·4
Σ	96·1	96·3

Tab. 9 lists the results of determining the composition of rocks at Venera-13 and Venera-14 landing sites.

It is evident from the Table that the compositions of rocks at Venera-13 and Venera-14 landing sites differ, thus reflecting different geomorphological provinces.

The upland rolling plain on which Venera-13 landed is the most widespread type of the Venusian surface. The composition of rock at Venera-13 landing site proved to be close to the composition of the potassium alkaline basalts of the Earth's crust. This type of rock is little spread on the terrestrial surface. It mainly occurs on oceanic islands and on continental rifts of the Mediterranean area. Apparently we met a similar type of rock at Venera-8 landing site: Venera-8 landed in a similar geomorphological province and detected the same potassium content as Venera-13 (see Tab. 4).

Venera-14 landed in the province typical of lowlands. The rock at the landing site turned out to be close in composition to tholeitic basalts comprising the Earth's oceanic crust, that is, close to the rocks which are widespread on the Earth.

It seems that in all where Veneras 8, 9, 10, 13 and 14 conducted investigations the rocks are close in composition to the basaltic rocks of the Earth's crust.

It is common knowledge that the abundance of water on the Earth has led to the great diversity of rocks. On Venus there is no such diversity of rocks apparently because of the lack of water there. Perhaps the great portion of the Venusian surface, including upland rolling plains and flat lowlands, is covered with basaltic rocks which outcrop from different depths and in different periods of the formation of the crust. Only rocks lying in the Ishtar and Aphrodite highlands remain unexplored. Structurally and morphologically they resemble the Earth's continents. Such are the general considerations concerning the data on the X-ray fluorescent analysis of Venusian rocks.

c) *X-ray fluorescent analysis of Martian rocks Investigations by Vikings 1 and 2*

The main objective of Vikings 1 and 2 experiments were to conduct exobiological studies and to detect the signs of the existence of life on Mars at present or in the past. One of the characteristics of the environment in which the search of the signs of life took place in it was its chemical composition. That is why the set of the scientific instruments of Vikings 1 and 2 included X-ray fluorescent spectrometers.

To determine the composition of Martian rocks use was made of the X-ray radiometric method of analysis and X-ray spectrometric instrumentation described in [11]. The rock sample for analysis was taken by the automatically controlled scoop and delivered to the measuring cavity of the spectrometer. Then it was irradiated by ^{55}Fe and ^{109}Cd radioisotopic sources. Excited fluorescent radiation was recorded by the X-ray fluorescent spectrometer.

Fig. 9 shows the scheme of the analysis and the location of the sample which was being measured and of radioisotopic sources and detectors — proportional counters. Two sources, four detectors and two calibration targets were located around the measuring cavity. The location

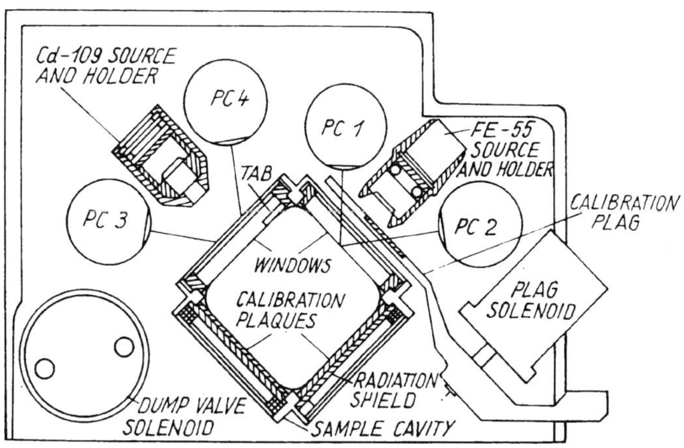

Fig. 9. Cross-sectional diagram of analysis section of X-ray instrument used on "Viking-1 and -2"

Tab. 10. The elemental composition of Martian soil samples*

Elements	Viking-1 sample 1	sample 2	sample 3	Viking-2 sample 1
Mg	5·0 ± 2·5	—	5·2	—
Al	3·0 ± 0·9	—	2·9	—
Si	20·0 ± 2·5	20·8	29·5	20·0
S	3·1 ± 0·5	3·8	3·8	2·6
Cl	0·7 ± 0·3	0·8	0·9	0·6
K	<0·25	<0·25	<0·25	<0·25
Ca	4·0 ± 0·8	3·8	4·0	3·6
Ti	0·51 ± 0·2	0·51	0·51	0·61
Fe	12·7 ± 2·0	12·6	13·1	14·2
O	50·1 ± 4·3	49·7	—	50·4
Rb	≦30			≦30
Sr	60 ± 30			100 ± 40
Y	70 ± 30			50 ± 30
Zr	30			30 ± 20

* The content of the elements from Mg to Fe is given in % and from Rb to Zr in 10^{-4}%.

of the detectors and sources was such that a pair of detectors recorded the fluorescent radiation excited by the ^{55}Fe source with the activity of 240 mc either in the calibration target situated opposite it (when there was no rock in the cavity) or in the rock sample when it was placed into the cavity. The other pair of detectors recorded fluorescent radiation excited by the ^{109}Cd source with the activity of 55 mC also in the calibration target or in the rock sample. Proportional counters specially designed for this instrument were used as detectors. The spectra of fluorescent radiation were recorded in 128 energy channels with the capacity of 2^5 pulses per channel. Accumulated information was kept in the memory of the onboard computer and transmitted at the request to Earth via Mars's artificial satellites.

The elemental composition of Martian soil samples measured by Vikings 1 and 2 is listed in Tab. 10.

The basic results of the experiments in an X-ray-radiometric analysis of Martian soil by Vikings 1 and 2 are the following:

— the composition of the fine-grained fraction of soil (grains are less than 2 mm in size) at both landing sites proved to be strikingly similar despite the fact that they lie 6500 km apart;

— the fine-grained fraction of soil consists mainly from one or more silicate minerals (the average content in samples is $SiO_2 = 45 \pm 5$ weight percent) with a high iron content (the average content in samples is $Fe_2O_3 = 19 \pm 3\%$);

— the sulfur content proved to be unexpectedly high — approximately 100 times higher than the average value for the Earth's crust. (Perhaps its main part forms magnesium sulfate);

— the potassium content is less than 0·25%, which is at least by 80 per cent lower than the average potassium content in the Earth's crust.

As a result of the above experiment the first data on the composition and characteristics of Martian soil were obtained. These data made it possible to get some ideas of the origin of Martian soil and the conditions of its shaping.

3. The main results of studying the Moon and the planets by nuclear-physics methods

We have briefly considered the results of using nuclear-physics methods of analyzing matter on some bodies of the solar system. Only gamma-spectrometry and X-ray spectrometry were considered, as the most widely used methods in space research.

These methods of analysis made it possible to study the chemical composition of the basic geomorphological provinces and the main types of the rocks forming the crust of the Moon and the planets. Natural radioactive elements were determined by the method of gamma-spectrometry. The content and ratios of these elements show the character and the origin of rocks. The basic rock-forming elements were determined by the X-ray spectrometry method, which enables one to classify the rocks and understand the conditions of their formation. New data obtained by the nuclear-physics method provide now the basis for the development of the ideas about the origin, evolution and modern structure of the bodies of the solar system.

Within the framework of this report the speaker is not supposed to analyze the geochemical and cosmochemical ideas which appeared on the basis of the results of the above-mentioned experiments. That is why we shall touch only upon one global consequence which follows from the analysis of the data on the composition of lunar, Venusian and Martian rocks.

Summing up the results of the consideration of the above data, one can say that on all celestial bodies a common geochemical process of the differentiation of matter takes place. This process devides them into envelopes. The upper envelope consists mainly of the rocks of the basic composition. On all bodies the bimodal geomorphological structure of the surface is shaped — ancient uplands and younger territories of a volcanogenic character, which are situated lower.

Ancient territories are formed in the earliest periods of the bodies' existence (4·5 billion years ago) as a result of their heating during the accretion. Apparently they represent the remnants of the ancient crust. Such a gabbro-anorthosite crust has survived now only on the Moon and perhaps on Mars and Mercury. Younger areas formed by basaltic rocks came into existence later (3—3·8 billion years ago) due to the tectonic and volcanic activity in the period of the heating of the upper mantle by radiogenic heat accumulated there.

On such little differentiated bodies as the Moon and Mars the compositions of the rocks of both provinces differ to an insignificant degree. Large planets, like the Earth (and perhaps Venus), passed through a deeper stage of the differentiation and, therefore, rocks considerably differing in composition lie on their surface. The difference in the character of the rocks lying on small and large bodies is due to the different degree of their differentiation which in its turn was predetermined by their primordial mass and composition.

Received October 25, 1983

References

[1] *A. P. Vinogradov, Yu. A. Surkov, G. M. Chernov* et al., Kosmich. Issled. (Cosmic Research) 5 (1967) 874

[2] *Yu. A. Surkov*. Gamma-Spectrometry in Space Research. Moscow: Atomizdat Publishers 1977

[3] *J. Trombka, J. R. Arnold, I. Adler* et al., Cosmochemistry of the Moon and the Planets. Mocow: Nauka Publishers 1975

[4] *G. H. Petengill, P. G. Ford, H. Masursky, E. Eliason*, Report to the 23rd Plenary Meeting, Budapest (Hungary) 1980

[5] *Yu. A. Surkov, L. P. Moskalyova, O. S. Manvelyan, A. T. Basilevsky, V. P. Kharyukova*, Proc. 11th Lunar Planet. Sci. Conf. **1980, 669**

[6] *Yu. A. Surkov*, Geokhimiya (Geochemistry) **1982**, 775

[7] *S. L. Mandelshtam, I. T. Tindo, V. I. Karev*, Kosmich. Issled. **VI** (1966) 827

[8] *G. Ye. Kocharov, S. V. Viktorov, V. I. Chesnokov*. The Nuclear Physical Investigations of the Moon and the Planets. Moscow: Energoizdat Publishers 1981

[9] *Yu. A. Surkov, B. M. Andreichikov, I. K. Akhmetshin* et al., Kosmich. Issled. **10** (1972) 930

[10] *I. Adler, J. Trombka, J. Gerard*, et al., Science **175** (1972) 536 and **177** (1972) 256

[11] *B. C. Clarke III, A. K. Baird, H. J. Rose, P. Toulmin III* et al., J. Geophys. Res. **82** (1977) 4577

[12] *Yu. A. Surkov, F. F. Kirnozov, V. N. Glazov* et al., Geokhimiya No. 1 (1981) 3

[13] *Yu. A. Surkov, O. P. Shcheglov, L. P. Moskalyova*, et al., Analit. Khimiya (Analyt. Chem.) **XXXVII** (1982) 1349

Isotopenpraxis **20** (1984) 9, S. 330 bis 339

Originalarbeiten

An Overall Theory of Separation on Column or on Square Cascade

I. Hodor

(Institute of Isotopic and Molecular Technology, Cluj-Napoca)[1])

An overall theory of separation on a column or on a square cascade is developed on the basis of only a few assumptions, irrespective of the particular structure of the separating element or the involved physico-chemical processes. The main application of the theory is that it clarifies and facilitates the derivation of the mass transfer parameters. It is illustrated by examples that well known theories as that of thermal diffusion column may be reproduced. The theory is also applied to spray column whose theory has been little developed so far.

Auf der Grundlage von nur einigen wenigen Annahmen und unabhängig von der speziellen Struktur des Trennelements sowie von den ablaufenden physikalisch-chemischen Prozessen wird eine allgemeine Theorie für die Trennung an einer Säule oder an einer Rechteck-Kaskade entwickelt. Der wichtigste Vorzug der Theorie besteht darin, daß sie zur Klärung und Erleichterung der Ableitung von Massentransport-Parametern beiträgt. Anhand von Beispielen wird gezeigt, daß bekannte Theorien, wie die der thermischen Diffusionssäule, reproduziert werden können. Die Theorie wird auch auf Rieseltürme angewandt, deren Theorie bisher wenig entwickelt war.

Keywords

column separation; isotope separation; mass transfer; separation processes; thermal diffusion

1. Introduction

A column or a square cascade is an equipment on which an elementary separation process is multiplied on a row of identical elements connected in series. The elements may be discrete as in a plate column, in a cascade of centrifuge, etc., or they may be continuous or differential as in a packed column, in a thermal diffusion column, etc. A complete description of the separation process would necessitate to establish the local equations for all implied phenomena and the adequate boundary conditions and then to solve the whole mathematical problem and to interpret the results. This procedure may have chance of success in some very simple cases but it is, generally, hopeless. For this reason, it is a current practice to introduce one or some overall transfer parameters whose deffinition may depend on the type of the equipment. In the case of a plate column such a parameter is the Murphree plate efficiency, in the case of packed column it may be the height of an equivalent theoretical plate (HETP). With these parameters the differential problem takes a simple form which may be handled easily.

A problem of a practical interest, to which the present paper refers, is to derive the mass transfer parameters from the local system of equations which is supposed to describe the mass transfer phenomena. Some examples are to be mentioned: *Lewis* [1] has derived for some cases the Murphree plate efficiency as function of point efficiency; *Kuhn* [2] and *Westhaver* [3] have derived the HETP of the hairpin-countercurrent; for thermal diffusion (TD) column, *Furry*, *Jones*, and *Onsager* [4], *Cohen* [5] and others have derived the coefficients which enter a simple overall differential equation. In each case, a particular reasoning, closely related to the concrete separation process, was developed and specific approximations were made.

The purpose of this paper is to show that, if some justified and not very restrictive postulates are made, then, the composition field, expressed, for example, by the mole fractions, has some features common to all columns or square cascades. These features are expressible mathematically and they depend neither of the particular structure of the separating element nor of the involved physical processes. The main application of this result is that it clarifies and facilitates the derivation of the mass transfer parameters. This application is illustrated by a few examples.

Some essential parts of this paper are comprised in [6].

2. The Mathematical Problem of the Differential Equipment. The Asymptotic Solution

Consider the steady state separation process of a binary mixture take place on the differential column sketched in Fig. 1. To fix the ideas, consider the composition of the mixture is expressed by the mole fraction of one of the components,

$$N = N(\tau, z)$$

where z is the longitudinal coordinate and τ denotes one or more transversal coordonates of a point in column.

All algebric, differential and integrodifferential equations which are to be satisfied by the field $N(\tau, z)$ can be divided in three groups, and each group may be written in a symbolical manner

$$\begin{cases} \boldsymbol{L}N(\tau, z) = 0\,, \qquad z_1 < z < z_2 & \text{(1a)} \\ \boldsymbol{L}_1 N(\tau, z_1) = 0 & \text{(1b)} \\ \boldsymbol{L}_2 N(\tau, z_2) = 0 & \text{(1c)} \end{cases}$$

[1]) Address: Cluj-Napoca 5, PO Box 700, Romania

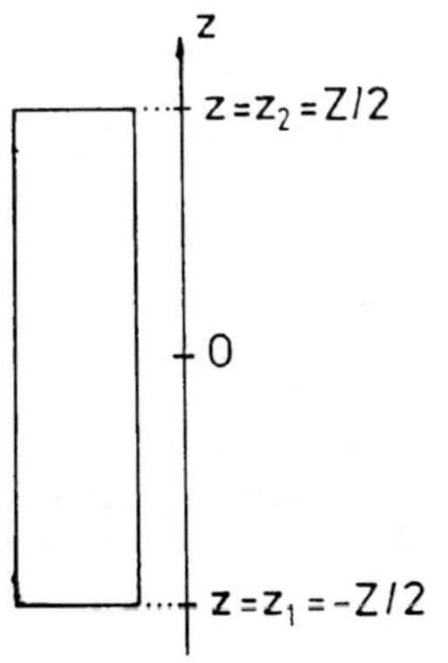

Fig. 1. Column with differential elements

The symbolical equation (1a) includes all equations which are to be satiesfied by $N(\tau, z)$ in inner points of the column. The equations (1b) and (1c) include all exterior conditions imposed to the field $N(\tau, z)$ at the two ends of the column. The system (1) is considered to be complete, e. g., it leads to a unique solution for the field $N(\tau, z)$.

The differential equipment is considered to be composed of identical elements connected in series. It means that the local equations (1a) do not depend explicitely on z. It follows that:

Postulate 1. *If $N(\tau, z)$ is a solution of the equation (1a) then $N(\tau, z + \Delta z)$, where Δz is an arbitrary quantity, is also a solution of Eq. (1a).*

A second basic assumption is that the differential equipment may be operated, theoretically, in a state of equilibrium in which the concentration field does not depend on z. If the process may be represented in a *McCabe-Thiele* diagram, the operating line of this state is reduced to a point on the equilibrium curve. Actually, every point on the equilibrium line represents such a state. It follows that:

Postulate 2. *The Eq. (1a) admits a particular solution $N_e = N_e(\tau, N_0)$ which does not depend on z and depends on an arbitrary constant N_0.*

It is useful to consider that the arbitrary constant is the concentration in a point $\tau = \tau_0$,

$$N_e(\tau_0, N_0) = N_0 \tag{2}$$

Generally, the system (1) is not linear in N which gives serious mathematical dificulties. For this reason it will be considered that the equipment is operated in such a way that the field $N(\tau, z)$ differs from an equilibrium field $N_e(\tau, N_0)$ only by a small perturbation.

At the ends of the column every point is considered to have contact with a separate infinite reservoir of constant concentration. Namely, at the end $z = z_1$, the point of coordinate τ has contact with a reservoir whose concentration is considered to be

$$N_e(\tau, N_0 + \delta_1) = N_e(\tau, N_0) + \frac{\partial N_e(\tau, N_0)}{\partial N_0}\,\delta_1\,,$$

$$z = z_1 = -Z/2 \tag{3}$$

and similarly, at the other end,

$$N_e(\tau, N_0 + \delta_2) = N_e(\tau, N_0) + \frac{\partial N_e(\tau, N_0)}{\partial N_0}\,\delta_2\,,$$

$$z = z_2 = Z/2 \tag{4}$$

where δ_1 and δ_2 are two very small quantities.

It is to note that the explicit form of the limit conditions (1b) and (1c), which form is not important here, might be obtained from (3) and (4) only if the operator $\boldsymbol{L}$ is known explicitly.

Generally, δ_1 and δ_2 could be arbitrary functions of τ in which case the system would have an infinity of freedom degrees. However, it is known from practice that at every end of the column only the mean value of the concentration is of interest. It follows that it is justified to leave for the system only two degrees of freedom, that is an unspecified mean concentration for every end of the column. Thus, it will be considered that δ_1 and δ_2 are two very small but arbitrary constants.

If $\delta_1 = \delta_2$, the solution of the system (1) is the equilibrium field $N_e = N_e(\tau, N_0 + \delta_1)$. If $\delta_1 \neq \delta_2$, the concentration field, may be written

$$N(\tau, z, N_0 + \delta_1, \delta, Z) = N_e(\tau, N_0 + \delta_1) + \frac{\partial N}{\partial \delta}(\tau, z, N_0, 0, Z)\,\delta \tag{5}$$

where the higher order terms were neglected and

$$\delta = \delta_2 - \delta_1 \tag{6}$$

Intuitively it is clear that:

Postulate 3. *If the legnth of the column increases unlimited, $Z \to \infty$, the concentration field tends to an equilibrium field.*

It follows that

$$\lim_{Z\to\infty} \frac{\partial N}{\partial \delta}(\tau, z, N_0, 0, Z) = a\,\frac{\partial N_e}{\partial N_0}(\tau, N_0) \tag{7}$$

where a is a constant.

Near the ends, the field is strongly influenced by the limit conditions. In the points which are not near the ends, the field is completely developed and this form of the field, denoted by N_∞, has the most important role in the separation process.

Consider the limit

$$\lim_{Z\to\infty} \frac{1}{M}\left[\frac{\partial N}{\partial \delta}(\tau, z, N_0, 0, Z) - a\,\frac{\partial N_e}{\partial N_0}(\tau, N_0)\right] = n(\tau, z, N_0) \tag{8}$$

where

$$M = M(Z, N_0) = \max\left|\frac{\partial N}{\partial \delta}(\tau, z, N_0, 0, Z) - a\,\frac{\partial N_e}{\partial N_0}(\tau, N_0)\right| \tag{9}$$

Now, if the length of the column increases, the field in the central zone of the column, in its tendency to the equilibrium field, has the asymptotical form

$$N_\infty = N_e(\tau, N_0 + \delta_1) + a\,\frac{\partial N_e}{\partial N_0}(\tau, N_0)\,\delta + n(\tau, z, N_0)\,M\delta = N_e(\tau, N_0 + \delta_1 + a\delta) + n(\tau, z, N_0)\,M\delta$$

Including the constant $\delta_1 + a\delta$ in the constant N_0, the field N_∞ may be written

$$N_\infty = N_e(\tau, N_0) + \beta n(\tau, z, N_0) \tag{10}$$

where $\beta = M\delta$ is a new arbitrary small constant. Now, N_0 has the role of an arbitrary constant which is finite, not small. The relation (10) shows how N_∞ depends of the two degrees of freedom N_0 and β.

The field (10) has to satisfy the equation (1a), but, according to postulate 1, the field

$$N^*_\infty = N_e(\tau, N_0) + \beta n(\tau, z + \Delta z, N_0) \quad (11)$$

must also satisfy Eq. (1a). N^*_∞ is also a fully developed field. The increase Δz may be interpreted as an arbitrary constant, thus the field (11) seams to depend on three arbitrary constants although the system has only two degrees of freedom. The contradiction disappears if, and only if, an increase Δz is equivalent to some increase ΔN_0 and/or $\Delta\beta$ of the constants N_0 and β, that is

$$N_e(\tau, N_0) + \beta n(\tau, z + \Delta z, N_0) = N_e(\tau, N_0 + \Delta N_0) + (\beta + \Delta\beta)\, n(\tau, z, N_0 + \Delta N_0) \quad (12)$$

Supposing that Δz is small and neglecting the second order terms one obtains

$$\frac{\partial n}{\partial z}(\tau, z, N_0) - \lambda n(\tau, z, N_0) = \frac{\Delta N_0}{\beta\, \Delta z}\frac{\partial N_e}{\partial N_0}(\tau, N_0) \quad (13)$$

where

$$\lambda = \frac{\Delta\beta}{\beta\, \Delta z} \quad (14)$$

The solution of Eq. (13) may be written in the form

$$n = \frac{1}{\lambda}\frac{\Delta N_0}{\beta\, \Delta z}\left[\psi(\tau, N_0, \lambda)\, e^{\lambda z} - \frac{\partial N_e}{\partial N_0}(\tau, N_0)\right] \quad (15)$$

The arbitrary function ψ has to be such a function that for $\lambda \to 0$, the solution (15) takes the form

$$n = \frac{\Delta N_0}{\beta\, \Delta z}\left[\psi_0(\tau, N_0) + \frac{\partial N_e}{\partial N_0}(\tau, N_0)\, z\right] \quad (16)$$

Indeed, this is the solution of Eq. (13) for $\lambda = 0$. It follows that

$$\psi(\tau, N_0, \lambda) = \frac{\partial N_e}{\partial N_0}(\tau, N_0) + \psi_0(\tau, N_0)\,\lambda + \frac{1}{2}\frac{\partial^2\psi}{\partial\lambda^2}(\tau, N_0, 0)\,\lambda^2 + \ldots \quad (17)$$

Introducing (15) into (10) one obtains

$$N_\infty = N_e(\tau, N_0) + \frac{1}{\lambda}\beta_0\left[\psi(\tau, N_0, \lambda)\, e^{\lambda z} - \frac{\partial N_e}{\partial N_0}(\tau, N_0)\right] \quad (18)$$

where

$$\beta_0 = \Delta N_0/\Delta z$$

Two of the increases ΔN_0, Δz and $\Delta\beta$ are arbitrary because they are connected only by the relation (12). λ cannot be an arbitrary constant because it is not in a position of arbitrary constant predicted by Eq. (10).

If (18) is compared with (10) one may write

$$\beta = \beta_0/\lambda \quad \text{for} \quad \lambda \neq 0 \quad (19)$$

Taking into account that

$$N_e(\tau, N_0) - \beta\frac{\partial N_e}{\partial N_0}(\tau, N_0) = N_e(\tau, N_0 - \beta) \quad (20)$$

and including β in the arbitrary constant N_0, the fully developed field of concentration is of the form

$$N_\infty = N_e(\tau, N_0) + \beta\psi(\tau, N_0, \lambda)\, e^{\lambda z} \quad \text{for} \quad \lambda \neq 0 \quad (21)$$

Similarly

$$N_\infty = N_e(\tau, N_0) + \beta_0\left[\psi_0(\tau, N_0) + \frac{\partial N_e}{\partial N_0}(\tau, N_0)\, z\right] \quad \text{for} \quad \lambda = 0 \quad (22)$$

The parameter λ and the functions $N_e(\tau, N_0)$, $\psi(\tau, N_0, \lambda)$ or $\psi_0(\tau, N_0)$ will be determined in every case by substituting N_∞ in the explicit form of Eq. (1a).

The field N in the central zone of the equipment tends to N_∞ if the column lengthens and the concentrations given at the two ends are only slightly derivated from an equilibrium field. For this reason N_∞ may be called the asymptotic field.

In practice, offen only the longitudinal variation of the concentration is of interest so that it is useful to introduce the mean transversal concentration defined by

$$\bar{N}_\infty = \bar{N}_\infty(z) = \int_\sigma g(\tau)\, N_\infty(\tau, z)\, d\tau \quad (23)$$

where σ is the transversal section of the equipment, $d\tau$ is the "surface" element and $g(\tau)$ is a bounded unspecified weight function with the property

$$\int_\sigma g(\tau)\, d\tau = 1 \quad (24)$$

In a given case, $g(\tau)$ will be chosen on the basis of some practical arguments so that $\bar{N}_\infty$ has some useful physical meaning.

If Eq. (21) is integrated to obtain mean values, the resulting equation is

$$\bar{N}_\infty(z) = \bar{N}_e + \beta\bar{\psi}\, e^{\lambda z}$$

The arbitrary constants $\bar{N}_e$ and β can be eliminated to obtain

$$\frac{d^2\bar{N}_\infty}{dz^2} = \lambda\frac{d\bar{N}_\infty}{dz} \quad (25)$$

which is a general differential equation for the transversal mean concentration. Note that (25) is satisfied for the field of the form (22).

3. The Equipment with Discrete Elements

The formal theory presented above is easily extended to this type of equipment which is presented schematically in Fig. 2.

It is considered that each element has its own three-dimensional coordinate system and the three coordonates of a point are denotes by a single symbol τ^*. The concentration field is thus written as

$$N = N(s, \tau^*)$$

where s is the serial number of an element or stage. Similar points in different elements have the same spatial coornate τ^* but different values for s, two different points in an element have different values for τ^* and the some value for s.

The equivalent equation to (1a) is

$$LN(s, \tau) = 0\,, \qquad s_1 \leqq s \leqq s_2 \quad (26)$$

that is, the continuous coordinate z is replaced by the discrete coordonate s. The symbolical Eq. (26) includes both the equations which are to be satisfied by N in inner points of the element and the connection relations between elements.

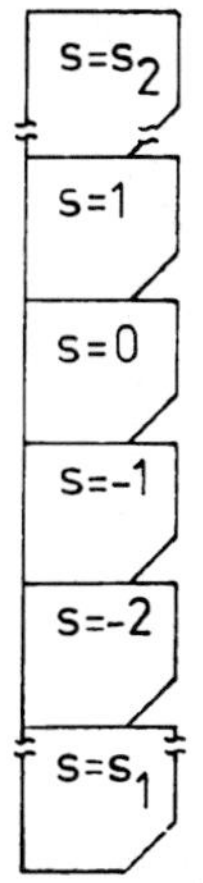

Fig. 2. Column with discrete elements

The postulate 1 is to be substituted by

Postulate 1*. *If $N(s, \tau^*)$ is a solution of Eq. (26) then $N(s+1, \tau^*)$ is also a solution of Eq. (26).*

The relation corresponding to (10) is

$$N_\infty = N_e(\tau^*, N_0) + \beta n(s, \tau^*, N_0) \tag{27}$$

and that corresponding to (12) is

$$\begin{aligned} &N_e(\tau^*, N_0) + \beta n(s+1, \tau^*, N_0) \\ &\quad = N_e(\tau^*, N_0 + \Delta N_0) + (\beta + \Delta\beta) \\ &\quad \times n(s, \tau^*, N_0 + \Delta N_0) \end{aligned}$$

Finally, the form of the fully developed concentration field corresponding to (21) and (22) is

$$N_\infty = N_e(\tau, N_0) + \beta\psi(\tau, N_0, \lambda)\, e^{\lambda s} \quad \text{for} \quad \lambda \neq 0 \tag{28}$$

$$N_\infty = N_e(\tau, N_0) + \beta_0 \left[\psi_0(\tau, N_0) + \frac{\partial N_e}{\partial N_0}(\tau, N_0)\, s\right] \quad \text{for} \quad \lambda = 0 \tag{29}$$

4. Other Mathematical Presentations of the Theory

The I lines which interconnect the elements in the equipment presented schematically in Fig. 3 are some ways on which mass is transported, not necessarily by convection. Each way i is supposed to be characterised by an only one concentration and let be denoted by $N(i, s)$ the concentration corresponding to the way i between the elements s and $s+1$. The way number $i = 1, 2, ..., I$ may be regarded as a discrete transversal coordonate so that

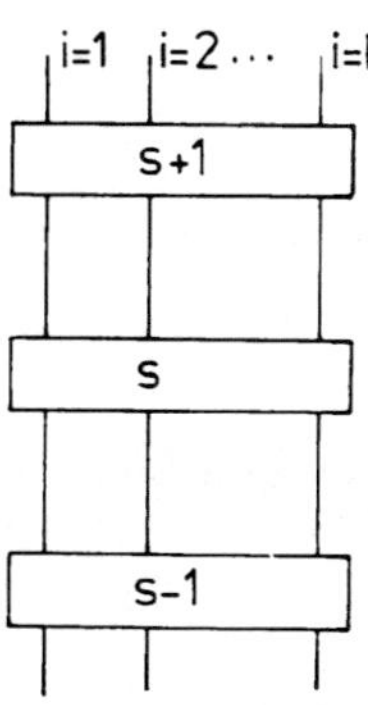

Fig. 3. Column for which both longitudinal and transversal coordinates are discrete

the two types of solution of the infinit column may be written as

$$N_\infty = N_e(i, N_0) + \beta\psi(i, N_0, \lambda)\, e^{\lambda s} \quad \text{for} \quad \lambda \neq 0 \tag{30}$$

$$N_\infty = N_e(i, N_0) + \beta_0 \left[\psi_0(i, N_0) + \frac{\partial N_e}{\partial N_0}(i, N_0)\, s\right] \quad \text{for} \quad \lambda = 0 \tag{31}$$

It may be also possible that transversal variable is discrete but the longitudinal one is continuous. This is the case when some streams are circulated along the column and between streams a mass exchange, characterised by some transfer coefficients, takes place. At a given longitudinal coordinate, each stream is characterised by an only one value of concentration. In such a case, the discrete coordinate s in (30) and (31) will be substituted by the continuous coordinate z.

5. The Multicomponent Case

The theory may be easily extended to the multicomponent case. Only the differential equipment analised in Section 2 will be presented here in a schematical manner.

Suppose there are ν components which means that the concentrations of $\nu - 1$ components are independent. The equipment may be operated, theoretically, at the equilibrium, in which case the independent concentrations of the $\nu - 1$ components are equilibrium concentrations, indendenpent of z,

$$N_{ei} = N_{ei}(\tau_0, N_{01}, N_{02}, ..., N_{0\nu-1}), \quad i = 1, 2, ..., \nu - 1 \tag{32}$$

where, similar to (2),

$$N_{ei}(\tau_0, N_{01}, N_{02}, ..., N_{0\nu-1}) = N_{0i} \tag{33}$$

N_{0i} being arbitrary concentrations.

Now, consider that each independent field of concentrations is only very slightly deviated from equilibrium state. The effects which result from the interaction between the small deviations will be of the second order and can be neglected. It follows that each small deviation will follow the same laws as in the bicomponent case. That is, the fully developed concentration fields are of the same form as (21) and (22),

$$N_{i\infty} = N_{ei}(\tau, N_{0j}) + \beta_i\psi_i(\tau, N_{0j}, \lambda_i)\, e^{\lambda_i z} \quad \text{if} \quad \lambda \neq 0 \tag{34}$$

$$N_{i\infty} = N_{ei}(\tau, N_{0j}) + \beta_{0i}\left[\psi_{0i}(\tau, N_{0j}) + \frac{\partial N_{ei}}{\partial N_{0j}}(\tau, N_{0j})\, z\right] \quad \text{if} \quad \lambda_i = 0 \tag{35}$$

where N_{0j} means $N_{01}, N_{02}, ..., N_{0\nu-1}$. Note that there are $\nu - 1$ finite arbitrary constants N_{0j} and $\nu - 1$ arbitrary, infinitely small constants β_i.

6. A General Equation for the Close-Separation Equipment

One of the most interesting applications of the above theory is that it permits to establish a general equation for the close-separation column or cascade, that is, for those cases in which the elementary separation effect is small. The isotope separations enter in this cathegory.

Only the differential equipment for binary mixture is considered, but the result may be generalised to the multicomponent mixtures.

Note that (22) may be written in the form

$$N_\infty = N_e(\tau, N_0 + \beta_0 z) + \beta_0 \psi_0(\tau, N_0) \tag{36}$$

It means that the two fields N_∞ and N_e are parallel functions of z, the 'distance' $\beta_0\psi_0(\tau, N_0)$ between them being indenpendent of z. If one analyses examples of processes which may be represented in a *McCabe-Thiel* diagram, one finds out that the system (1) is satisfied by a solution of the type (22) if, and only if, the operating line is parallel to the equilibrium line [6].

It is useful to note that although (22) is strictly valid only if $\lambda = 0$, it is applicable also for λ very small. Indeed, if λ is very small (a small inifite), it follows from (18) that the field differs from (22) only with terms proportional to $\beta\lambda$ or smaller. But $\beta\lambda$ is a small inifite of the second order because β is also a small inifite. Hence, the solution (22) is applicable if N_∞ is parallel or almost parallel to N_e or, with orders words, if the angle between operating- and equilibrium line is zero or small. This condition is always satisfied in practice for a close separation column or cascade, therefore (22) is the pertinent solution for studying this type of equipment.

In close-separation equipment, the small elementary separation effect is multiplied, usually, very many times so that the concentration difference, at the two ends of the equipment, may be apreciable. But (22) is established in the hypothesis that the concentration difference, at the two ends, is very small. Still, it may be imagined that the equipment is extremely long and may be divided in long pieces, for each pice all the conditions, for which (22) was established, being satisfied. Hence each piece will be characterised by a field of the type (22) but the constants N_0 and β_0 may have different values from a column piece to another. For this reason, it is not practical to use (22) directly to describe the separation process for the whole column. It is useful to establish an overall differential equation and this is made subsequently.

If elementary separation effect is small, the equilibrium field may always be written in the form

$$N_e(\tau, N_0) = N_0 + \varepsilon^* N_e^*(\tau, N_0) \tag{37}$$

where ε^* is a small constant regarded, mathematically, as a small infinite of the first order. The function $N_e^*(\tau, N_0)$ is null for $N_0 = 0$ and for $N_0 = 1$, it is null also for azeotropic points. The equation (37) may be regarded as the definition relation of the close separation equipment.

Taking into account that β_0 is also a small inifite, in the case of close-separation equipment, the Eq. (22) may be written

$$N_\infty = N_e(\tau, N_0) + \beta_0[\psi_0(\tau, N_0) + z] \tag{38}$$

Suppose the equipment is defined so that N_∞ depends only on the two parameters $\boldsymbol{N_0}$ and β (or β_0). A function which depends on the field N_∞ will be also determined by the same parameters. z-component of the transport is such a function

$$J_z = J_z(N_0, \beta) \tag{39}$$

J_z does not depend on z because the field N_∞ is stationary and the components are conserved. It follows that a translation of z-axis has no influence upon J_z. Such a translation is equivalent in (21) by an increas of β and in (22) or (36) by an increase of N_0. It means that

$$J_z(N_0, \beta) = J_z(N_0, \beta + \Delta\beta) \quad \text{if} \quad \lambda \neq 0 \tag{40}$$

$$J_z(N_0, \beta_0) = J_z(N_0 + \Delta N_0, \beta_0) \quad \text{if} \quad \lambda = 0 \tag{41}$$

or

$$\frac{\partial J_z}{\partial \beta}(N_0, \beta) = 0 \quad \text{if} \quad \lambda \neq 0 \tag{42}$$

$$\frac{\partial J_z}{\partial N_0}(N_0, \beta_0) = 0 \quad \text{if} \quad \lambda = 0 \tag{43}$$

Suppose the z-component of the flux is of the form

$$j_z = j_{ze}(N_\infty, \tau) - cD\frac{\partial N_\infty}{\partial z} \tag{44}$$

where $j_{ze}(N_\infty, \tau)$ is an unspecified function, x is the total molar density of the two components and D is the coefficient of diffusion or of eddy-diffusion. c and D may depend on τ. Note that $j_{ze}(N_e, \tau)$ represents the flux for an equilibrium field. The expression (44) is sufficiently general, covering all practical situations. If still j_z has other expression, the theory may be developed in a way similar to that which follows.

Introducing N_∞ given by (22) into (44) and taking into account that β_0 is small one obtains

$$j_z = j_{ze}(N_e, \tau) + \frac{\partial j_{ze}}{\partial N_e}(N_e, \tau) \times \beta_0\left[\psi_0(\tau, N_0) + \frac{\partial N_e}{\partial N_0}(\tau_0, N_0)\, z\right] - cD\beta_0 \tag{45}$$

This expression is, now, introduced into

$$J_z = \int_\sigma j_z \, d\tau \tag{46}$$

It is obtained

$$J_z(N_0) = J_{ze}(N_0) + \beta_0\left[\int_\sigma \psi_0(I, N_0) \times \frac{\partial j_{ze}(N_e, \tau)}{\partial N_e} d\tau - \int_\sigma cD \, d\tau\right] \tag{47}$$

where

$$J_{ze}(N_0) = \int_\sigma j_{ze}(N_e, \tau)\, d\tau \tag{48}$$

is the transport for the equilibrium field $N_e(\tau, N_0)$. The term

$$\beta_0 \int_\sigma \frac{\partial j_{ze}(N_e, \tau)}{\partial N_e} \frac{\partial N_e(\tau, N_0)}{\partial N_0} d\tau = \beta_0 \int_\sigma \frac{\partial j_{ze} N_e(\tau, N_0)}{\partial N_0} d\tau = \beta_0 \frac{\partial J_{ze}(N_0)}{\partial N_0}$$

was necglected because it is a small infinite of the second order. Indeed, β_0 is a small infinite and

$$\frac{\partial J_{ze}(N_0)}{\partial N_0} = \frac{\partial J_z}{\partial N_0}(N_0, 0)$$

is zero if (43) is satisfied, otherwise it is a small infinite of the order of λ. Actually, it is implied here the natural supposition that $\partial J_z/\partial N_0$ is a continuous function of λ.

From (38)

$$\beta_0 = \frac{\partial N_\infty}{\partial z} = \frac{\partial N_\infty^0}{\partial z} \tag{49}$$

where $N_\infty = N^0_\infty(\tau_0, N_0)$ so that (47) may be written

$$J_z = J_{ze}(N_0) + \left[\int_\sigma \psi_0(\tau, N_0)\,\frac{\partial j_{ze}(N_e, \tau)}{\partial N_e}\,\mathrm{d}\tau - \int_\sigma cD\,\mathrm{d}\tau\right]\frac{\mathrm{d}N^0}{\mathrm{d}z} \qquad (50)$$

This is the general equation of the fine separation of a binary mixture on a column or on a square cascade. How it may be used it will be illustrated in the following section. Let be considered the plane case of a TD column.

7. The Thermal Diffusion (TD) Column

A fluid binary mixture is disposed between two vertical parallel walls, one cold, the other heated, Fig. 4. The flux of a species is

$$\boldsymbol{j} = cN\boldsymbol{v} + cD[-\operatorname{grad} N + \alpha_T N(1 - N)\operatorname{grad}\ln T] \qquad (51)$$

where $\boldsymbol{v}$ is the mean molar velocity; T is the field of the absolute temperature; α_T is the TD constant; D is the coefficient of ordinary diffusion and may depend on x only.

The transfer of momentum, energy and mass takes place simultaneously in a point of the column and is, generally, interdependent. In the case of an isotopic mixture, the influence of the field N upon the other fields c, $\boldsymbol{v}$, D and T is negligible. As a consequence it is allowed to suppose that hydrodynamic and energy problems were solved previously, that is, the fields c, $\boldsymbol{v}$, D and T will be considered as given functions of x only, $\boldsymbol{v}$ having only a z-component $v(x)$.

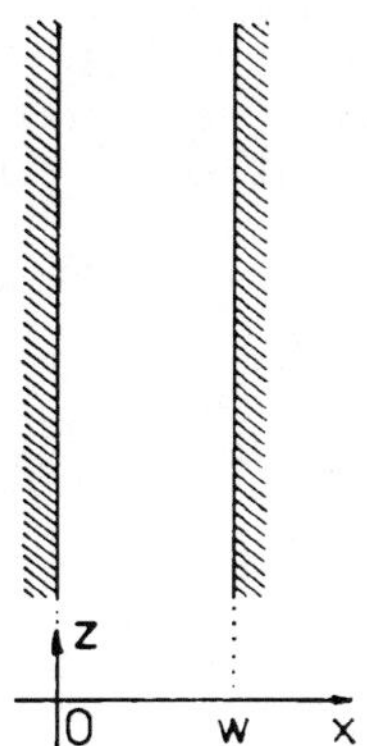

Fig. 4. Plane case of TD column

The process is supposed to be stationary, $\operatorname{div}\boldsymbol{j} = 0$, and this condition together with the boundary conditions at the two walls give the explicit form of the Eq. (1a),

$$LN(\tau, z) = 0$$

$$\Leftrightarrow \begin{cases} \dfrac{\partial}{\partial x}\left\{cD\left[-\dfrac{\partial N}{\partial x} + \alpha_T N(1 - N)\dfrac{\mathrm{d}\ln T}{\mathrm{d}x}\right]\right\} + cv\,\dfrac{\partial N}{\partial z} - cD\,\dfrac{\partial^2 N}{\partial z^2} = 0\,, \quad 0 > x > w & (52\mathrm{a}) \\ \dfrac{\partial N}{\partial x} + \alpha_T N(1 - N)\,\dfrac{\mathrm{d}\ln T}{\mathrm{d}x} = 0\,, \quad x = 0\,, \quad x = w & (52\mathrm{b}) \end{cases}$$

where $\Leftrightarrow$ is the symbol of logical equivalence; x has the rôle of τ, the implied fields being independent of the other transversal coordinate.

The system (52) satisfies the three postulates. The equilibrium concentration field N_e, independent of z, sitisfying the system (52) is given by

$$\frac{N_e(1 - N_0)}{N_0(1 - N_e)} = \exp\left(\alpha_T \ln\frac{T}{T_0}\right)$$

where N_0 and T_0 are the concentration and the temperature in a point $x = x_0$. For isotopic mixture, α_T is regarded as a small infinite of the first order so that

$$N_e = N_0 + \alpha_T N_0(1 - N_0)\ln(T/T_0) \qquad (53)$$

Note that α_T corresponds to ε^* in (37).

The z-component of the flux

$$j_z = cvN - cD\,\frac{\partial N}{\partial z} \qquad (54)$$

gives, for the equilibrium field (53),

$$j_{ze} = cvN_e - cvN_0 + cvN_0(1 - N_0)\,\alpha_T\ln(T/T_0) \qquad (55)$$

The equilibrium transport is

$$J_{ze} = PN_0 + BN_0(1 - N_0)\,\alpha_T\int_0^w cv\ln(T/T_0)\,\mathrm{d}x \qquad (56)$$

where

$$P = B\int_0^w cv\,\mathrm{d}x \qquad (57)$$

is the net molar flow in the positive z-direction; B is the breadth and w the thickness of the column. The derivative

$$\frac{\partial J_{ze}}{\partial N_0} = P + B(1 - 2N_0)\,\alpha_T\int_0^w cv\ln(T/T_0)\,\mathrm{d}x \qquad (58)$$

is zero if (43) is satisfied, otherwise it is a small infinite of the order of λ which means that P must be regarded as a small infinite.

To determine ψ_0, the solution of te form (22) is introduced into (52) in which case the derivative (58) is zero. Taking into account that β_0 and α_T are small infinites, one obtains

$$\frac{\mathrm{d}\psi_0}{\mathrm{d}x} = \frac{1}{cD}\int_0^x cv\,\mathrm{d}p \qquad (59)$$

Because ψ_0 is defined upto an arbitrary term of the form (arbitrary const.) $\partial N_e/\partial N_0$, it is sufficient to take into consideration a particular solution of Eq. (59) of the form

$$\psi_0 = \int_0^x \frac{\mathrm{d}q}{cD}\int_0^q cv\,\mathrm{d}P + 1 \qquad (60)$$

so that if $\tau_0 = x_0 = 0$ then $\psi_0 = 1$.

The equation (50) is a relation between small infinites. Indeed, the first term on the right hand of the equality is given by (56) where P and α_T are small inifnites. The second term is proportional with $\mathrm{d}N^0_\infty/\mathrm{d}z$ which, according to (49), is proportional to the small infinite β_0. It means that in calculating the brackets in (50), the small infinites may be neglected. One may write

$$\int_{\sigma} \varphi_0(\tau, N_0) \frac{\partial j_{ze}(N_e, \tau)}{\partial N_e} \mathrm{d}\tau = B \int_0^w cv \, \mathrm{d}x$$

$$\times \left(\left(\int_0^x \frac{\mathrm{d}q}{cD} \int_0^q cv \, \mathrm{d}p + 1 \right)^2 = P \left(1 + \int_0^w \frac{\mathrm{d}p}{cD} \int_0^x cv \, \mathrm{d}q \right) \right.$$

$$- B \int_0^w \frac{1}{cD} \left(\left(\int_0^x cv \, \mathrm{d}q^2 \right) \mathrm{d}x \cong - B \int_0^w \frac{1}{cD} \left(\int_0^x cv \, \mathrm{d}q \right)^2 \mathrm{d}x \right. \tag{61}$$

where an integration by parts was made.

The transversal variation of N_∞ is of the order of α_T which means that some transversal mean value of N_∞ differs from N_∞^0 only by a small infinite. It follows that N_∞^0 and N_0 in (50) may be substituted by some mean value $\overline{N}_\infty$.

Now the Eq. (47) may be written

$$J_z = P\overline{N}_\infty + N\overline{H}_\infty(1 - \overline{N}_\infty) - (K_c + K_d) \frac{\mathrm{d}\overline{N}_\infty}{\mathrm{d}z} \tag{62}$$

where, for the coefficients, were introduced the usual notations from the theory of the TD column:

$$H = B\alpha_T \int_0^w cv \ln(T/T_0) \, \mathrm{d}x \tag{63a}$$

$$K_c = B \int_0^w \frac{1}{cD} \left(\int_0^x cv \, \mathrm{d}q \right)^2 \mathrm{d}x \tag{63b}$$

$$K_d = \int_{\sigma} cD \, \mathrm{d}\tau = B \int_0^w cD \, \mathrm{d}x \tag{63c}$$

The Eq. (62) is well known in the theory of TD column. There are only minor differences between the expressions of the coefficients given by (63) and those given elsewhere [7]. It has been preferred here to use x as integrating variable instead of the temperature T. Another difference is produced by the fact that, in the literature, the concentrations and fluxes are expressed in different ways.

The theory presented in this paper may be rewritten for any mode of expression of the concentrations and fluxes. But in a given practical case a mode of expression for which the postulates of the theory are better fulfilled will give better results.

In the literature ([7] p. 36, [8] p. 101), there is also an equation of the TD column which has two additional terms

$$J_z = P\overline{N}_\infty + H\overline{N}_\infty(1 - \overline{N}_\infty) - (K_c + K_d) \times \frac{\partial \overline{N}_\infty}{\partial z} + K_t \frac{\partial \overline{N}_\infty}{\partial t} - K_{d'} \frac{\partial^2 \overline{N}_\infty}{\partial z^2} \tag{64}$$

where K_t and $K_{d'}$ are coefficients for which integral expressons similar to (63) were obtained. Numerical estimations have shown that the two suplimentary terms are small and may be neglected.

In the stationary case, $\overline{N}_\infty$ does not depend on t, and using (25) one may write

$$-K_{d'} \frac{\mathrm{d}^2 \overline{N}_\infty}{\mathrm{d}z^2} = -K_{d'} \lambda \frac{\mathrm{d}\overline{N}_\infty}{\mathrm{d}z}$$

Since λ and $\mathrm{d}\overline{N}_\infty/\mathrm{d}z$ are small infinites it follows that the last term in (64) is, theoretically, of the second order. It is our opinion that in deriving Eq. (64) some other terms of the second order were neglected so that an equation of the type (64) which contains all second order terms has not been derived so far.

8. Spray Column for Isotope Separation

A spray column is a simple vertical tube, without any interior structure as packing or plates. Two phases are circulated countercurrently, one of which being dispersed in the other in the form of small droplets. To fix the ideas we may have in mind the column used by *Lewis* and *Mac Donald* [9] for isotope separation: a liquid solution moves up through a tube and a fine spray of liquid amalgam moves down. Between the two phases an equiatomic isotope exchange takes place and this exchange is accompanied by a small elementary enrichment effect ε. The theory of spray column is little developed, therefore it is an appropriate example to show how the above theory may be used in such cases.

The spray column for isotope separation is described by the usual separation equation of the same type as (62),

$$J_z = P\overline{N}_\infty + \varepsilon L\overline{N}_\infty(1 - \overline{N}_\infty) - hL \frac{\mathrm{d}\overline{N}_\infty}{\mathrm{d}z} \tag{65}$$

where ε is considered to be small and known, h is the HETP. The Eq. (65) may be derived in a similar way as (62). But often the main theoretical problem for isotope separation column is to determine h from the local system of equations.

ε being small, the operating and equilibrium lines are parallel or almost parallel. h depends neither on isotope concentration nor on net molar flow P. For this reason, to derive h it may be supposed that the concentration N is very small in which case the mathematical problem is simpler.

Suppose now that the local system of equations for small concentrations ($N \ll 1$) is

$$\boldsymbol{L}N(\tau, z) = 0$$

$$\Leftrightarrow \begin{cases} D \dfrac{1}{r^2} \dfrac{\partial}{\partial r} \left[r^2 \dfrac{\partial X}{\partial r}(r, z) \right] = v \dfrac{\partial X(r, z)}{\partial z}, \quad 0 < r \leqq r_0 & \text{(66a)} \\ \lim\limits_{r \to 0} \dfrac{\partial X}{\partial r}(r, z) = 0 & \text{(66b)} \\ Dc_1 \dfrac{\partial X}{\partial r}(r_0, z) = k[\alpha Y(z) - X(r_0, z)] & \text{(66c)} \\ G \dfrac{\mathrm{d}Y}{\mathrm{d}z}(z) + D_e c_2 \dfrac{\mathrm{d}^2 Y}{\mathrm{d}z^2}(z) = \dfrac{3L}{c_1 v r} k[\alpha Y(z) - X(r_0, z)] & \text{(66d)} \end{cases}$$

Here, r is the radial coordonate in a droplet, and r_0 is the droplet radius. The continuous phase is considered completely mixed transversally, its isotop concentration $Y(z)$ being a function of z only. X is the field of concentration in a droplet and is supposed to have spherical symmetry. The transversal coordonate τ has a continuous domain $\tau = r \in [0, r_0]$ and has also a discrete value, say $\tau = \tau_0$, which represents any point in the continuous phase. Thus, the concentration field is

$$N(\tau, z) = \begin{cases} X(r, z) & \text{for} \quad \tau = r \in [0, r_0] \\ Y(z) & \text{for} \quad \tau = \tau_0 \end{cases}$$

Eq. (66a) is the equation of unsteady diffusion in spherical coordonate with spherical symmetry but the time is replaced by $t = z/v$, v being the velocity of the droplet. D is the coefficient of diffusion in the droplet. The sense of z-axis is from up to down.

Eq. (66a) includes the isotope conservation law but it does not contain the point $r = 0$. For this reason, the limit condition (65b) tels that the isotopes are not created or distroyed in the centre of the droplets.

To obtain Eq. (66c), it is imagined that the droplet carries a thin layer of the continuous phase which has negligible isotopic content but it puts up resistance to the itosoptic exchange. The left-hand member represents the isotopic flux at the inner side of the droplet surface. The right-hand member is the *Cohen*'s flux [10] through the interface for small concentrations. $\alpha = 1 + \varepsilon$ is the elementary separation factor. The transfer coefficient k is inversely proportional to the thickness of the thin layer carried by the droplet.

c_1 and c_2 are the total molar (atomic) densities in the dispersed and continuous phase respectively.

The z-component of transport in the continuous phase is supposed to have two terms

$$-GY(z) - \sigma D_e c_2 \frac{dY}{dz}(z) \tag{67}$$

The first term is due to convective movement, G being the total molar (atomic) rate of the continuous phase. The second term is due to eddy-diffusion; D_e is the coefficient of eddy-diffusion, σ is the mean cross section of the continuous phase.

If (67) is derived by z and the sign is changed, one obtain the left-hand member of Eq. (66d) which represents transversal isotopic flux to the other phase. The right-hand member of Eq. (66d) represents the same flux but it is obtained by multiplying the transport through the whole surface of one droplet

$$4\pi r_0^2 k[\alpha Y(z) - X(r_0, z)]$$

by the number of droplets per unit length of the column

$$\frac{L}{vc}\frac{3}{4\pi r^3} = \frac{3L}{4\pi c_1 r_0^3 v}$$

where L is the total molecular rate of the dispersed phase, $4\pi r_0^3/3$ is the volume of the droplet.

For small concentrations, the Eq. (65) takes the form

$$J_z = (P + \varepsilon L)\, \overline{N}_\infty - hL \frac{d\overline{N}_\infty}{dz} \tag{68}$$

If the field is of the form (22) the calculations are simplified the most. In this case, the last term in (68) does not depend on z. Because J_z does not depend on z too, it follows that the field (22) is possible if

$$P + \alpha L = 0 \tag{69}$$

For the system (66), the equilibrium field is

$$N_e = N_e(\tau, N_0) \Leftrightarrow \begin{cases} Y_e = N_0\,, & \tau = \tau_0 \\ X_e = \alpha N_0\,, & \tau = r \in [0, r_0] \end{cases} \tag{70}$$

so that the field of the type (22) is

$$Y_\infty = N_0 + \beta_0[\psi_0(\tau_0) + z] \tag{71a}$$

$$X_\infty = \alpha N_0 + \beta_0[\psi_0(r) + \alpha z] \tag{71b}$$

If (71b) is substituted in (66a) and (66b) one obtains

$$\psi_0(r) = (\alpha v/6D)\, r^2 \tag{72}$$

Here, an arbitrary term was neglected which is possible because $\psi_0(\tau)$ is determined upto an arbitrary term of the form (arbitrary const.) $\partial N_e/\partial N_0$. Now, if (66c) is used one obtaines

$$\psi_0(\tau_0) = \frac{c_1 v r_0}{3k\alpha} + \frac{v r_0^2}{6D} \tag{73}$$

so that the field N_∞ is

$$Y_\infty = N_0 + \beta_0\left(\frac{c_1 v r_0}{3k\alpha} + \frac{v r_0^2}{6D} + z\right), \qquad \tau = \tau_0 \tag{74a}$$

$$X_\infty = \alpha N_0 + \alpha\beta_0\left(\frac{v}{6D} r^2 + z\right), \qquad \tau = r \in [0, r_0] \tag{74b}$$

The Eq. (66d) was not used so far. If (74) is substituted in (66d) one obtain

$$G = \alpha L \tag{75}$$

which is equivalent to (69) because $P = L - G$. Thus the linear field (74) is a solution of the system (66) if and only if (75) is satisfied, that is, if the equilibrium line and operating line are parallel.

Now, the isotopic transfer along the column is composed of (67) and the transport made by the dispersed phase movement

$$J_z = L\overline{X}_\infty - GY_\infty - D_e c_2 \frac{dY_\infty}{dz} \tag{76}$$

where $\overline{X}_\infty$ is the mean concentration in a droplet

$$\overline{X}_\infty = \frac{3}{r_0^3}\int_0^{r_0} X_\infty r^2\, dr = \alpha N_0 + \alpha\beta_0\left(\frac{v r_0^2}{10D} + z\right) \tag{77}$$

Using (74), (76) gives

$$J_z = -\alpha\beta_0 L\left(\frac{v r_0^2}{15D} + \frac{c_1 v r_0}{3k\alpha} + \frac{\sigma D_e c_2}{\alpha L}\right) \tag{78}$$

It is to note that this expression of the transport is valid just if $\alpha - 1$ is not small but (75) is fulfilled.

In the case under discussion, the Eq. (23) is to be written

$$\overline{N}_\infty = g(\tau_0)\, Y_\infty(z) + \int_0^{r_0} g(r)\, X_\infty(r, z)\, dr$$

The weight function may be chosen as

$$g(\varepsilon) = \begin{cases} 1\,, & \tau = \tau_0 \\ 0, & \tau = r \in [0, r_0] \end{cases}$$

so that

$$\overline{N}_\infty = Y_\infty(z) = N_\infty + \beta_0\left(\frac{c_1 v r_0}{3k\alpha} + \frac{v r_0^2}{6D} + z\right) \tag{79}$$

For small $\varepsilon = \alpha - 1$, J_z may be calculated by introducing (79) in (68) and taking into account (69). It is obtained

$$J_z = -hL\frac{d\overline{N}_\infty}{dz} = -hL\beta_0 \tag{80}$$

If this is compared with (78) and the approximation $\alpha = 1$ is made, one obtains

$$h = \frac{v r_0^2}{15D} + \frac{c_1 v r_0}{3k} + \frac{\sigma D_e c}{L} \tag{81}$$

The first term is due to diffusion in the droplet, the second is due to the resistance oposed by the layer carried by the droplet, and the third term is due to longitudinal eddy-diffusion which takes place in the continuous phase.

9. The Plate Column

As a last illustration, consider the *Lewis'* case II [1]: A countercurrent vapour-liquid exchange takes place on a rectangular plate column sketched in Fig. 5. The vapour rises from plate to plate without mixing and the liquid flows in the same direction on all plates. It will be assumed that the vapour leaving a given plate at any point rises directly to corresponding point on the plate above. For simplification it will also be assumed that the local Murphree efficiency is $E = 1$, that is, the vapour leaving a plate is in equilibrium with the liquid in every point. It is not assumed that the elementary separation effect is small.

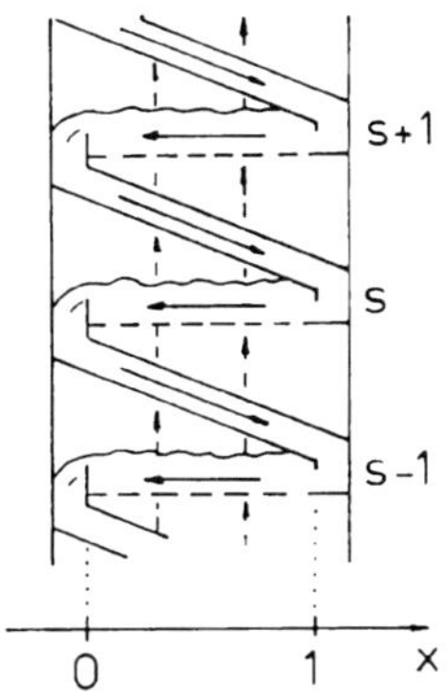

Fig. 5. Plate column

Continuous arows represent liquid flow.
Discontinuous arrows represent vapour flow.

The separation elements are discrete so that the equations are of the type (23),

$$\boldsymbol{L}N(s, \tau^*) = 0$$

$$\Leftrightarrow \begin{cases} R \dfrac{\mathrm{d}X_s(x)}{\mathrm{d}x} + Y_s(x) - Y_{s-1}(x) = 0 & (82\,\mathrm{a}) \\ Y_s(x) = f_e[X_s(x)] & (82\,\mathrm{b}) \\ X_s(1) = X_{s-1}(0) & (82\,\mathrm{c}) \end{cases}$$

Here R is the slope of the operating line, $X_s(x)$ is the mole fraction of high boiler in liquid on the plate s at the point x, $Y_s(x)$ is the mole fraction of the same component in the vapour phase, s is the plate number which regarded, in the theory, as the longitudinal discrete coordinate. Eq. (82 b) is the equation of the equilibrium curve.

In this case the symbolical coordinate τ^* may be considered as being composed of the continuous coordonate x and of the discrete coordonate i which takes only two values: 1 for liquid phase and 2 for the vapour phase. The concentration field is

$$N(\tau^*, s) = N(x, i, s) = \begin{cases} X_s(x)\,, & i = 1 \\ Y_s(x)\,, & i = 2 \end{cases}$$

Of cours, for any constant value $X_s = N_0$, Eq. (82 b) gives an equilibrium field

$$N_e = N_e(\tau^*, N_0) \Leftrightarrow \begin{cases} X_{se} = N_0 \\ Y_{se} = f_e(N_0) \end{cases} \tag{83}$$

which satisfies the whole system (82).

For a small vicinity of the field (83), the equilibrium Eq. (82 b) may be linearised

$$Y_s = K(N_0)\, X_s + b(N_0) \tag{82 b'}$$

Now, the solution of the form (28) may be written

$$X_\infty = N_0 + \beta\psi(x, N_0, \lambda)\, e^{\lambda s} \tag{84}$$

Substituting Y_s given by (82 b') and $X_s = X_\infty$ given by (84) in (82 a) and (82 c), it is obtained

$$\begin{cases} R \dfrac{\mathrm{d}\psi(x, N_0, \lambda)}{\mathrm{d}x} + (1 - e^{-\lambda})\, K(N_0)\, \psi(x, N_0, \lambda) = 0 & (85\,\mathrm{a}) \\ \psi(1, N_0, \lambda) = \psi(0, N_0, \lambda)\, e^{-\lambda} & (85\,\mathrm{b}) \end{cases}$$

which gives the solution

$$\psi = A \exp\left[\frac{K(N_0)\,(e^{-\lambda} - 1)}{R}\, x\right] \tag{86}$$

where λ is the solution of the characteristic equation

$$e^{-\lambda} = 1 - \frac{R}{K}\lambda \tag{87}$$

Using this result one may obtain the overall plate efficiency. It is to note that the result is the same as that obtained by *Lewis*.

10. Final Remarks

The theory presented in this paper is based on two principal approximations: one consists in the linearisation of the mathematical problem in the vicinity of an equilibrium state, the other consists in considering that the column or cascade is very long and in neglecting the end-effects. The end-effects mean the perturbations of the concentration field near the column ends. With other words, this theory describes the asymptotic behaviour of the column when its state approaches an equilibrium state and, at the same time, the length of the column increases indefinitely in both senses. One could believe that these approximations are too rough. In fact, they are normally accepted in the literature, more or less implicitly. The theory of various isotopic separation processes on column, as TD, distilation, centrifugation, electromigration, etc., are based on some approximations which are equivalent to those made here. These approximations are accepted even if the elementary separation effect is not small. For example, in Sec. 9 it is illustrated how the *Lewis'* results may be reproduced. Just the acceptance to use an only one transfer parameter, as HETP, to describe an intricate exchange phenomenon on column is implicitly based on the two approximations.

It is very profitable to use the present theory to an unstudied type of column which is illustrated in Sec. 8.

In a given case it may happen to exist more, just an infinite number of solutions of the type (21). The solutions differ from one another by different values for λ. The true asymptotic field corresponds to that λ which has the minimum absolute value. The appearance of more solutions of the form

$$\beta\psi(\tau, N_0, \lambda)\, e^{\lambda z}$$

is explained by the fact that they participate to describe the end-phenomena.

The Eq. (87) has at the most one real solution different of zero. But it has an infinite number of complex solutions which will participate to describe the end-effects.

The end effects for a particular case are considered in [11]. It is note to that the special mathematical aspects which appear in [11] were studied by *Beals* [12].

Received June 17, 1983

References

[1] *W. K. Lewis*, Ind. Eng. Chem. **28** (1936) 399
[2] *W. Kuhn*, Helv. Chim. Acta **25** (1942) 252
[3] *J. W. Westhaver*, Res. Nat. Bur. Stand. **38** (1947) 169
[4] *W. H. Furry, R. C. Jones, L. Onsager*, Phys. rev. **55** (1939) 1083
[5] *K. Cohen*, Proc. of Int. Symp. on Isotope Sep., Amsterdam 1957, North-Holland Publ. Co. Amsterdam 1958, p. 3
[6] *I. Hodor*, Thesis (univ. Babeş-Bolyai, Cluj-Napoca, Romania, 1971)
[7] *Gh. Văsaru, G. Müller, G. Reinhold, T. Fodor*, Thermal Diffusion Column. Berlin: VEB Deutscher Verlag der Wissenschaften 1969
[8] *P. H. Abelson, N. Rosen, J. I. Hoover* (Eds.), Liquid Thermal Diffusion. USAEC/TID-5229 1958
[9] *G. N. Lewis, R. T. MacDonald*, J. Amer. Chem. Soc. **58** (1936) 2519
[10] *K. Cohen*, "Theory of Isotope Separation", New York: McGraw Hill 1951
[11] *I. Hodor* (inder press in Studia Univ. Babeş-Bolyai, Phys.)
[12] *R. Beals*, J. Math. Phys. **22** (1981) 954

Isotopenpraxis **20** (1984) 9, S. 339 bis 344

Бинарная экстракция*

А. И. Холькин, В. И. Кузьмин

(Sibirische Abt. der AdW der UdSSR, Inst. für Chemie u. chem. Technologie, Krasnojarsk)[1]

Бинарная экстракция имеет ряд свойств совместно с координативной экстракцией, а также особенности, связанные со свойствами экстрагентов, работающих на основе обмена анионами или катионами, а различается от этих систем не только в описании равновесия экстакции. В связи с этим является целесообразным выделить их как новый класс экстракционных процессов. Описана экстракция кислот, солей и гидроокислов металлов, рассматриваются примеры и приведены экстракционные ряды. Благодаря многочисленным возможностям вариации бинарного экстрагента существует ряд вариантов влияния, например на активность, и тем самым управления разделительными процессами. Таким образом, бинарная экстракция ведёт к новым эффективным решениям гидрометаллургических заданий.

The binary extraction is characterized by a number of properties being also typical for the coordinative extraction as well as by peculiarities in respect to the properties of anion and cation exchanging extractants. Therefore it is suitable to consider it a new class of extraction processes. The extraction of acids, salts and metal hydroxides is described, examples are discussed, and extraction series are given. Due to the manifold possibilities of the variation of the binary extractant there are a number of variants to influence e. g. selectivity, and thus to control the processes of separation. The binary extraction thus leads to new efficient solutions of hydrometallurgical problems.

Keywords

inorganic acids; liquid column chromatography; organic solvents; separation processes; solvent extraction

К бинарной экстракции относятся процессы извлечения кислот, солей и гидроокисей металлов солями органических кислот и органических оснований. Исследование таких систем позволило выявить ряд их существенных особенностей и выделить их в новый класс экстракционных процессов.

Впервые работы по исследованию экстракции солями органических кислот и органических оснований были проведены Гринстедом с сотр. [1, 2]. Однако они не получили должного развития, так как в них не было дано широкого анализа закономерностей экстракции и, как следствие, не были выявлены необычайно широкие возможности применения процесса бинарной экстракции для решения задач гидрометаллургии.

Нами [3, 4] было показано, что системы с бинарными экстрагентами характеризуются рядом свойств, общих с координационной экстракцией, имеют особенности, связанные со свойствами анионообменных и катионообменных экстрагентов, и в то же время, отличительные черты, присущие только этим системам, проявляющиеся как в описании экстракционных равновесий, так и в новых возможностях управления процессом.

Рассмотрим процесс бинарной экстракции минеральных кислот для солей четвертичных аммониевых оснований и органических кислот

$$mH^+_{(b)} + B^{m-}_{(b)} + mR_4NA_{(0)} \rightleftarrows (R_4N)_mB_{(0)} + mHA_{(0)} \tag{1}$$

для солей аминов

$$mH^+_{(b)} + B^{m-}_{(b)} + mR_3NHA_{(0)} \rightleftarrows (R_3NH)_mB_{(0)} + mHA_{(0)} \tag{2}$$

При сравнении его с процессом извлечения кислоты нейтральными экстрагентами, например, аминами

$$mH^+_{(b)} + B^{m-}_{(b)} + mR_3N_{(0)} \rightleftarrows (R_3N)_mH_mB_{(0)} \tag{3}$$

следует, что и в том и в другом случае в органическую фазу переходят молекулы H_mB или, что точнее,

---

* Доклад был представлен на 3. Семинаре по экстракции Центрального института исследований материалов, ГДР — Тюрмсдорьф, 27. 3–31. 3. 1983 г.

[1]) Адрес: 660097 Krasnojarsk 97, UdSSR

катионы и анионы в стехиометрическом отношении, то есть характер распределения кислот для этих двух классов процессов одинаков. В то же время в основе процессов лежат разные типы химических реакций: при бинарной экстракции — реакция ионного обмена, при экстракции нейтральными экстрагентами — реакция присоединения. Эти сходства и отличия проявляются на характере распределения веществ в этих системах и, соответственно, отражаются в уравнениях, описывающих процессы.

Запишем выражения для констант равновесия обоих процессов: для бинарной экстракции

$$K_{HmB} = \frac{C_{(R_4N)_mB_{(0)}} \cdot C^m_{HA_{(0)}}}{C^m_{R_4NA_{(0)}} \cdot A^m_{H^+_{(b)}} \cdot C_{B_b^{m-}}} \cdot \frac{\gamma_{(R_4N)_mB_{(0)}} \cdot \gamma^m_{HA_{(0)}}}{\gamma^m_{R_4NA_{(0)}} \cdot \gamma_{B^{m-}_{(b)}}} \tag{4}$$

для извлечения нейтральным экстрагентом

$$K_{HmB} = \frac{C_{(R_3N)_mHmB_{(0)}}}{C^m_{R_3N_{(0)}} \cdot A^m_{H^+_{(b)}} \cdot C_{B^{m-}_{(b)}}} \cdot \frac{\gamma_{(R_3N)_mHmB_{(0)}}}{\gamma^m_{R_3N_{(0)}} \cdot \gamma_{B^{m-}_{(b)}}} \tag{5}$$

Из уравнений (1—5) видно, что на извлечение аниона можно влиять изменением активности ионов водорода и наоборот — на распределение ионов водорода влиять изменением концентрации аниона в водной фазе.

Взаимная зависимость распределения противоионов друг от друга объединяет бинарную экстракцию и экстракцию нейтральными экстрагентами и отличают их от катионо- и анионообменных процессов.

Нетрудно показать, что при условии постоянства состава органической фазы, то есть при экстракции малых количеств кислоты HmB, в зависимости коэффициентов распределения аниона B^{m-} от кислотности водной фазы однотипны как для бинарной экстракции, так и извлечения нейтральным экстрагентом.

$$\lg D_B = \text{const} - m\,pH \tag{6}$$

На рис. 1 приведены данные по экстракции бромистоводородной кислоты в зависимости от pH водной фазы раствором триоктиламина (1), ди-2-этилгексилфосфатами триоктиламина (2) и тетраоктиламмония (3). Можно видеть, что для обоих типов экстрагентов влияние кислотности на экстрагируемость аниона идентична: зависимости линейны и имеют одинаковый угол наклона.

На следующем рис. 2 приведены аналогичные данные по бинарной экстракции минеральных кислот различной основности растворами н-трет. бутилфенолята тетраоктиламмония. Полученные зависимости также линейны и тангенс угла наклона прямых равен

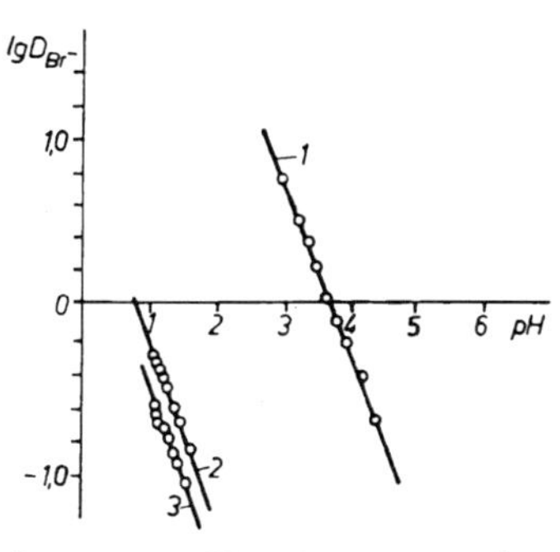

Рис. 1. Экстракция HBr нейтральным и бинарным экстрагентами

1 – триоктиламин 0,1 М в толуоле; 2 – ди(2-этилгексил)фосфат триоктиламина 0,1 М, ди(2-этилгексил)фосфорная кислота 0,3 М в толуоле; 3 – ди(2-этилгекзил)фосфат тетраоктиламмония 0,1 М, ди(2-этилгексил)фосфорная кислота 0,3 М в толуоле

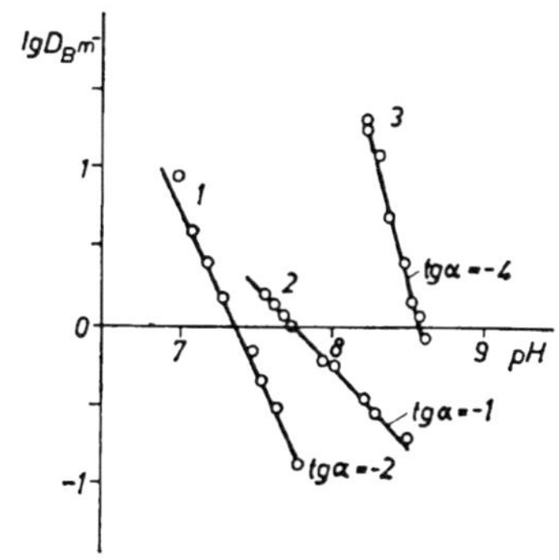

Рис. 2. Экстракция H_2SO_4 (1), HCl (2) и $H_4Fe(CN)_6$ (3)

0,1 М раствором n-третбутилфенолята тетраоктиламмония в толуоле в присутствии 0,3 М n-третбутилфенола для извлечения HCl и H_2SO_4 и 0,2 М для извлечения $H_4Fe(CN)_6$. $C_{Cl^-_{(исх)}} = 1$ моль/л, $C_{SO^{2-}_{4\,(исх)}} = 0{,}01$ моль/л, $C_{Fe(CN)^{4-}_{6(исх)}} = 0{,}0025$ моль/л

заряду экстрагируемых анионов в соответствии с уравнением (6).

Таким образом, соли органических кислот и органических оснований имеют общие свойства с нейтральными экстрагентами. В результате извлечения в органической фазе образуются катионо- и анионообменный экстрагент, каждый из которых в отдельности этих свойств не проявляет.

Для примера, показывающего отличие экстракционных систем с бинарными экстрагентами от систем с нейтральными экстрагентами, рассмотрим характер изотерм при экстракции кислот.

При стехиометрии протонов и анионов минеральной кислоты в водной фазе (отсутствие высаливателей) и в органической фазе ($mC_{HA_{(0)}} = C_{(R_4N)_mB_{(0)}}$) и постоянстве коэффициентов активности компонентов в органической фазе изотерма распределения кислоты согласно уравнению (4) описывается уравнением:

$$C_{HmB_{(0)}} = \bar{K}_{HmB}^{\frac{1}{m+1}} \cdot C_{R_4NA}^{\frac{m}{m+1}} \cdot C_{HmB_{(b)}} \cdot \gamma_{\pm} \tag{7}$$

Аналогичное выражениедл я извлечения кислоты нейтральными экстрагентами для процесса

$$mH^+_{(b)} + B^{m-}_{(b)} + pL_{(0)} \rightleftarrows HmB \cdot pL_{(0)} \tag{8}$$

имеет вид:

$$C_{HmB_{(0)}} = \bar{K}_{HmB} \cdot C^p_{(0)} \cdot C^{m+1}_{HmB_{(b)}} \cdot \gamma_{\pm}^{m+1} \tag{9}$$

Оба уравнения (7) и (9) близки по форме, но имеют существенное отличие, заключающееся в наличии показателей m = I для сомножителя $C_{HmB_{(b)}} \cdot \gamma_+$ в последнем случае.

Если изотерма при бинарной экстракции имеет приямолинейный характер ($C_{R_4NA_{(b)}} = \text{const}$), то при экстракции нейтральными экстрагентами зависимости криволинейны, причем характер их зависит от основности экстрагируемых кислот. Эти отличия продемонстрированы на рис. 3, на котором приведены изотермы экстракции соляной кислоты бинарным экстрагентом — ди-2-этилгексилдитиофосфатом тетраоктиламмония (рис 3а), и нейтральным — три-н-бутилфосфатом (б).

Выразим константу бинарной экстракции через константы более простых процессов. Это позволяет связать экстрагируемость кислот (далее солей и гидроокисей) с физико-химическими характеристиками исходных систем с органическими кислотами и систем

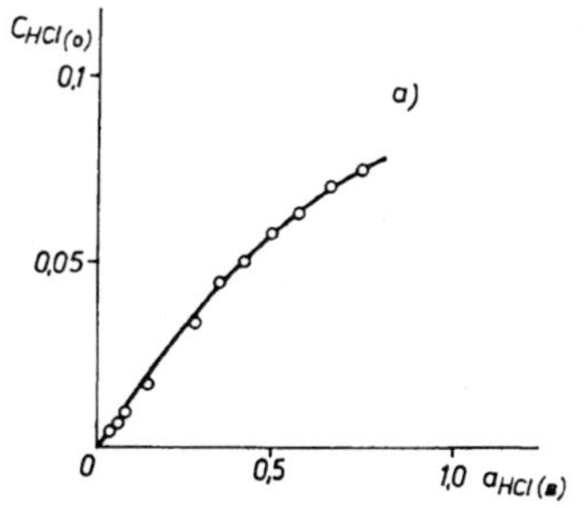

Рис. 3(а). Изотерма экстракции HCl ди(2-этилгексил)-дитиофосфатом тетраоктиламмония (0,1 М в толуоле)

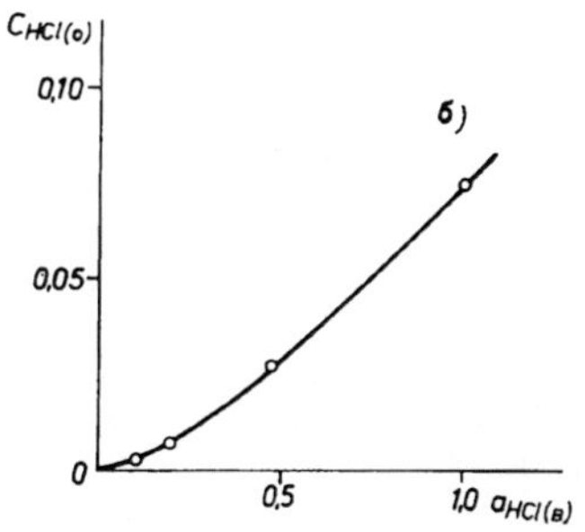

Рис. 3(б). Изотерма экстракции HCl трибутилфосфатом

с минеральными солями органических оснований, которые в настоящее время достаточно хорошо изучены.

Принимая, что в водной фазе соединения HmB, R_4NA и $(R_4N)_mB$ полностью диссоциированы, находим выражение для константы бинарной экстракции

$$K_{HmB} = \frac{K_{(R_4N)_mB} \cdot K_{HA}^m}{K_{R_4NA}^m} \cdot \frac{1}{K_a^m} \tag{10}$$

где $K_{(R_4N)_mB}$, K_{R_4NA} и K_{HA} — константы распределения электролитов $(R_4N)_mB$, R_4NA и нейтральной кислоты, а K_a — константа кислотной диссоциации органической кислоты.

Для извлечения кислоты нейтральными экстрагентами можно получить

$$K_{HmB} = \frac{K_{HmB \cdot pH}}{K_L^p} \cdot K_b^m \tag{11}$$

где K_L и $K_{HmB \cdot pH}$ — константы распределения экстрагента и экстрагируемого соединения, а K_b — константа основности экстрагента.

Из уравнения (11) следует, что экстракционная способность нейтральных экстрагентов при извлечении одной и той же кислоты HmB, во многом определяется основностью экстрагента (Kb), то-есть константой процесса

$$L + H^+ \rightleftarrows LH^+ \tag{12}$$

Очевидно, что экстракция кислот бинарными экстрагентами будет существенно зависеть от величин I/Ka, то-есть будет определяться процессом

$$A^- + H^+ \rightleftarrows HA \tag{13}$$

Экспериментальные данные по извлечению минеральных кислот солями тетраоктиламмония и органических кислот, константа кислотной диссоциации которых изменялась в широких пределах, подтверждают это заключение. На рис. 4 приведены данные по бинарной экстракции соляной кислоты. Как видно из рисунка, экстракционная способность экстрагентов увеличивается в ряду диалкилдитиофосфат < диалкилфосфат < каприлат < n-третбутилфенолят ЧАО. Этот порядок коррелирует с уменьшением величины константы кислотной диссоциации исходных органических кислот.

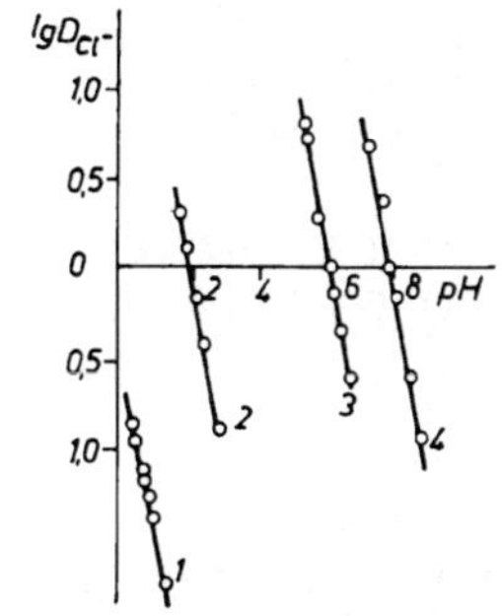

Рис. 4. Влияние природы органической кислоты на экстракцию HCl 0,5M растворами солей органических кислот тетраоктиламмония

1 — ди(2-этилгексил)дитиофосфат ЧАО, 0,5 М ди(2-этилгексил)дитиофосфорная кислота; 2 — ди(2-этилгексил)фосфат ЧАО, 0,5 М ди(2-этилгексилфосфорная кислота; 3 — n-каприлат ЧАО, 0,5 М каприловая кислота; 4 — n-третбутилфенолят ЧАО, 1,5 М n-третбутилфенол

Значительный интерес представляет рассмотрение процессов разделения минеральных кислот в системах с бинарными экстрагентами, то есть рассмотрение их экстракционной способности по отношению к различным минеральным кислотам.

Выразим константу бинарной экстракции через константы других элементарных процессов, протекающих в системе: константы обмена анионов минеральной и органической с гидроксилионом в системах с солями ЧАО — $K_{B\text{-}OH}$ и $K_{A\text{-}OH}$

$$K_{HmB} = \frac{K_{HA}^m \cdot K_{B\text{-}OH}}{K_a^m \cdot K_{A\text{-}OH}^m} \tag{14}$$

Из этого уравнения следует, что при использовании одного и того же бинарного экстрагента (значения K_{HA}, K_a и $K_{A\text{-}OH}$ одинаковы) экстрагируемость (K_{HmB}) будет определяться величиной $K_{B\text{-}OH}$, т.е. кислоты будут экстрагироваться тем лучше, чем лучше экстрагируется их анион в системах с солями ЧАО. Для примера, на рис. 5 даны результаты экстракции кислот n-третбутилфенолятом тетраоктиламмония. Из рисунка видно, что экстрагируемость падает в ряду HJ > HBr > HCl. Эти данные соответствуют известному ряду экстрагируемости для солей ЧАО.

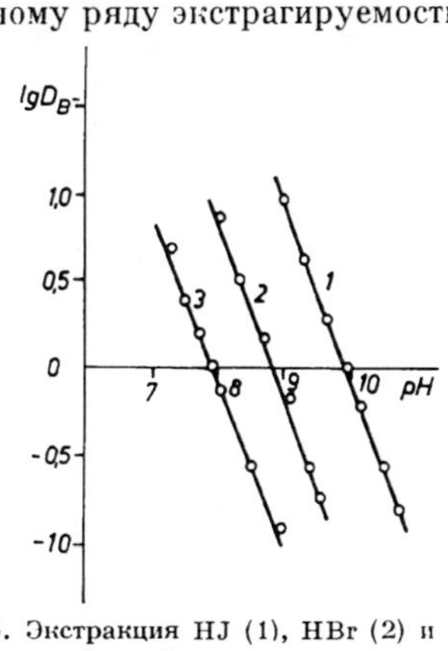

Рис. 5. Экстракция HJ (1), HBr (2) и HCl (3) 0,5 М раствором n-третбутилфенолята тетраоктиламмония в толуоле в присутствии 1,5 М n-третбутилфенола

Исходные концентрации J^-, Br^-, Cl^- — 0,05 моль/л

Следует обратить внимание на высокие коэффициенты распределения кислот при высоких pH, то есть в щелочной области. Бинарные экстрагенты на основе алкилфенолов оказались настолько эффективными, что экстрагируютья минеральные кислоты из нейтральных и щелочных растворов. При этом водный раствор становится более щелочным. Такими свойствами не обладает ни один из известных нейтральных экстрагентов.

Аналогичным образом рассмотрим экстракцию солей металлов M_mB_n бинарными экстрагентами R_4NA

$$mM^{n+}_{(b)} + nB^{m-}_{(b)} + mnR_4NA_{(0)} \rightleftarrows n(R_4N)_mB_{(0)} + mMAn_{(0)} \quad (15)$$

По характеру межфазного распределения компонентов этот процесс аналогичен, также как и в предыдущем случае, извлечению солей нейтральными экстрагентами (координационной экстракции).

$$mM^{n+}_{(b)} + nB^{m-}_{(b)} + pL_{(0)} \rightleftarrows MmBn \cdot pL_{(0)} \quad (16)$$

Отличаются это два процесса типом химической реакции, протекающей в системах. При бинарной экстракции — это ионный обмен, при координационной экстракции — реакция присоединения.

По типу химической реакции этот процесс идентичен с анионо- и катионообменными процессами. Однако, характер межфазного распределения компонентов этих систем отличается. Поскольку бинарная экстракция по характеру распределения наиболее близка к координационной экстракции, далее мы будем сравнивать ее именно с этим процессом.

В соответствии с уравнениями (15, 16) константы процессов бинарной и координационной экстракции имеют вид соответственно:

$$K_{MmBn} = \frac{C^n_{(R_4N)_mB(0)} \cdot C^m_{MAn(0)}}{C^m_{M^{n+}_{(b)}} \cdot C^n_{B^{m-}_{(b)}} \cdot \gamma^{mn}_{R_4NA(0)}} \times \frac{\gamma^n_{(R_4N)_mB(0)} \cdot \gamma^m_{MAn(0)}}{\gamma^n_{B^{m-}_{(b)}} \cdot \gamma^m_{M^{n+}_{(b)}} \cdot \gamma^{mn}_{R_4NA(0)}} \quad (17)$$

$$K_{MmBn} = \frac{C_{MmBn} \cdot pL_{(0)}}{C^p_{L(0)} \cdot C^m_{M^{n+}_{(b)}} \cdot C^n_{B_{(b)}}} \cdot \frac{\gamma_{MmBn} \cdot pL_{(0)}}{\gamma^p_{L(0)} \cdot \gamma^m_{M^{n+}_{(b)}} \cdot \gamma^n_{B^{m-}_{(b)}}} \quad (18)$$

Как и при бинарной экстракции кислот, влияние состава водной фазы на экстракционное равновесие подобно для обоих типов процессов — экстрагируемость ионов зависит от активности противоионов, что не имеет место для межфазного ионного обмена. Согласно уравнению (17), в условиях постоянства коэффициентов активности компонентов логарифмы коэффициентов распределения катиона и аниона связаны между собой линейной зависимостью, причем тангенс угла наклона соответствует отношению зарядов катиона и аниона

$$\lg D_{Mn+} = \text{const} - \frac{n}{m} \lg D_{B^{m-}} \quad (19)$$

На рис. 6 приведены данные по распределению нитратов Ni и Co при бинарной экстракции солей ди(2-этилгексил)дитиофосфатом тетраоктиламмония, на рис. 7 — данные экстракции хлорида никеля каприлатом тетраоктиламмония. Полученные данные согласуются с уравнением (19).

Возможность влияния на коэффициенты распределения одного из ионов изменением концентрации

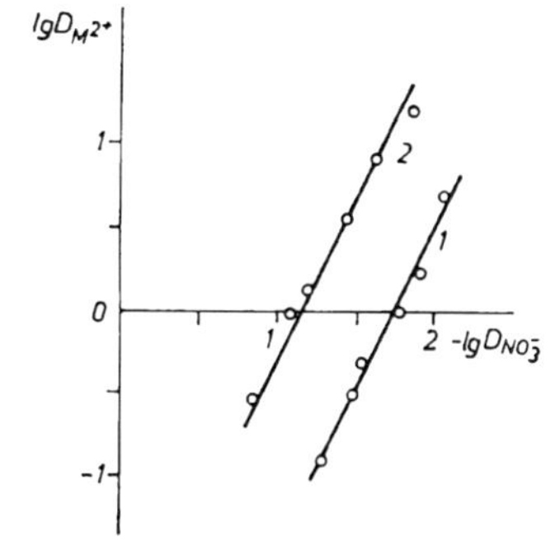

Рис. 6. Зависимость экстракции нитратов кобальта (1) и никеля (2) от концентрации нитрат-ионов при бинарной экстракции солей 0,25 M раствором ди(2-этилгексил)дитиофосфатом тетраоктиламмония в толуоле

$C_{Ni^{2+}_{(\text{исх})}}$ = 0,020 моль/л, $C_{Co^{2+}_{(\text{исх})}}$ = 0,0187 моль/л

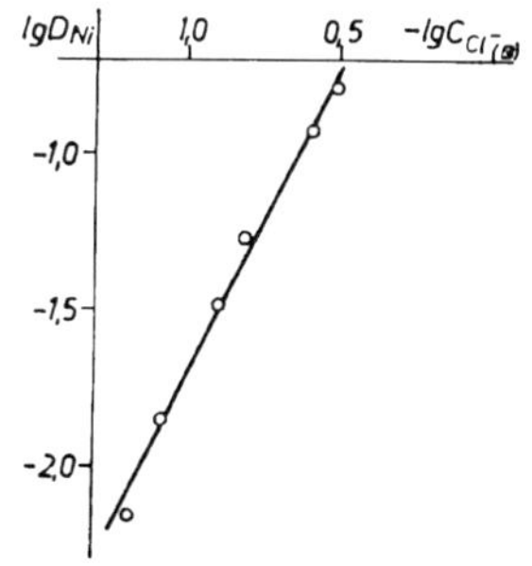

Рис. 7. Зависимость экстракции хлорида никеля от концентрации хлорид-ионов в водной фазе при бинарной экстракции 0,05 M раствором каприлата тетраоктиламмония в присутствии 0,05 M хлорида тетраоктиламмония

другого, противоположного по знаку, является важным свойством, характерным также для нейтральных экстрагенгентов.

Эта возможность широко используется в экстракционной технологии при извлечении металлов и очистке солей. В ряде случаев процесс очистки и концентрирования солей может быть организован без введения в систему каких-либо реагентов. Так например, при экстракции из сложных растворов, в присутствии высаливателей, можно достаточно эффективно извлекать соли металла в органическую фазу, а реэкстрагировать их водой.

Для бинарной экстракции возможности управления процессом еще более расширяются, поскольку эффект высаливателя применен также и для органической фазы, в которой можно изменять соотношение продуктов экстракции введением в систему органических солей металла или соответвстующих солей органического основания (см. уравнение 17). В этом проявляется одна из важных отличительных особенностей бинарной экстракции.

Другое отличие проявляется в описании изотерм бинарной и координационной экстракции. Согласно уравнениям (15) и (16) в условиях постоянства коэффициентов активности компонентов в органической фазе при отсутствии высаливателей выражение для изотермы бинарной экстракции записывается

$$\lg C_{MmBn_{(0)}} = \frac{1}{n \pm 1} \lg \overline{K}_{MmBn} + \frac{n}{n+1} \lg C_{R_4NA_{(0)}} + \lg (C_{MmBn_{(b)}} \gamma_{\pm}) \quad (20)$$

Mitteilungen / Information

IAEA: STRAHLENKONSERVIERUNG VON NAHRUNGSMITTELN

Eine internationale Beratungsgruppe fuer die Strahlenkonservierung von Nahrungsmitteln wurde am 9. Mai 1984 gegruendet. Mitglieder dieser Gruppe sind Aegypten, Argentinien, Bangladesh, die BRD, Frankreich, Irak, Israel, Kanada, die Niederlande, Mexiko, die Philippinen, Syrien, Thailand, Tuerkei und VR Ungarn. Interesse an der Teilnahme haben Finnland, Kostarika, Malawi und Portugal gezeigt. Verschiedene andere Mitgliedsstaaten werden der Gruppe wahrscheinlich in der naechsten Zukunft beitreten. Die Gruendung dieser internationalen Beratungsgruppe ist die Folge entscheidender Fortschritte auf dem Gebiet der Nahrungsmittelkonservierung durch Strahlung in den vergangenen 4 Jahren. Ein gemeinsames Experten-Kommittee der IAEA, der WHO und der FAO fuer die Unbedenklichkeit von strahlenkonservierten Nahrungsmitteln hat 1980 die Freigabe von Nahrungsmitteln empfohlen, die durch Strahlung mit einer Dosis von bis zu 10 Kilogray konserviert worden sind. Im Jahre 1983 hat die Codex Alimentarius Kommission allgemeine Richtlinien fuer strahlenkonservierte Nahrungsmittel angenommen. Diese Kommission wurde 1962 zur Schaffung eines gemeinsamen Richtlinienprogramms der WHO und FAO fuer Nahrungsmittel gegruendet. Diese Entwicklungen auf der internationalen Ebene und erweitertes Interesse an der Strahlenkonservierung von Nahrungsmitteln von seiten der Nahrungsmittelindustrie, der Handelsges. und von Regierungen haben die zustaendigen Stellen fuer die oeffentliche Gesundheit in vielen Mitgliedstaaten angeregt, neue oder zusaetzliche Regeln fuer die Bereitstellung von strahlenkonservierten Nahrungsmitteln fuer den menschlichen Konsum aufzustellen. Eine zunehmende Zahl von Staaten haben strahlenkonservierte Nahrungsmittel fuer den menschlichen Konsum freigegeben oder arbeiten gerade an Regeln zur Qualitaets- und Sicherheitsueberwachung. Im Jahre 1983 wurde an die Staaten, die Mitglied der IAEA, FAO oder WHO sind, eine Erklaerung verteilt, die Modalitaeten fuer die internationale Zusammenarbeit bei der Strahlenkonservierung von Nahrungsmitteln beschreibt und die Errichtung einer internationalen Beratungsgruppe wie die nun gegruendete vorschlug. Die Gruppe, die sich nun fuer eine Anlaufzeit von fuenf Jahren gebildet hat, trat formell in Kraft, nachdem 15 Staaten die Deklaration schriftlich angenommen hatten. Ihre Hauptziele bestehen darin, die weitere Entwicklung und kommerzielle Anwendung der Strahlenkonservierung von Nahrungsmitteln zu foerdern und Mitgliedstaaten - besonders Entwicklungslaender - bei der Verbesserung der Versorgung ihrer Bevoelkerung mit gesunden und ausreichenden Nahrungsmitteln zu unterstuetzen. Die IAEA, FAO und WHO haben ihre Bereitschft zur Unterstuetzung der Gruppe erklaert. (IAEA-Pressekommunique 84/13)

IAEA: ZIVILRECHTLICHE HAFTUNG BEI ATOMSCHAEDEN

Das Staendige Kommittee fuer zivilrechtliche Haftung bei Atomschaeden hielt im Gebaeude der IAEA in Wien in der ersten Maiwoche 1984 seine 5. Tagung ab. Das Kommittee wurde im Jahre 1963 von der IAEA gegruendet, um Probleme im Zusammenhang mit der am 21.5.1963 erarbeiteten Wiener Konvention ueber die zivilrechtliche Haftung bei Atomschaeden zu ueberpruefen. Diese Konvention trat am 12. November 1977 in Kraft. An der Tagung nahmen Vertreter und Beobachter von 26 Mitgliedstaaten der IAEA und der Nuklearenergie- Agentur der OECD (NEA) sowie von 2 Versicherungsgesellschaften teil. Das Staendige Kommittee fand keine wesentlichen Probleme hinsichtlich der Anwendung der Konvention oder der Nutzung ihrer Vorschlaege als Basis fuer die nationale Gesetzgebung in den Mitgliedstaaten der IAEA. Es ist eines der wesentlichen Ziele der Wiener Konvention, die Schadenshaftungshoehe zu definieren, die von Rechts wegen vom Betreiber von Nuklearanlagen gewaehrleistet werden muss. Die Konvention soll weiterhin sicherstellen, dass die notwendigen Geldmittel jederzeit verfuegbar sind, um Schadensersatzansprueche aus Nuklearschaeden zu erfuellen. Von seiten der Versicherungen wurde auf der Tagung unterstrichen, dass heute aufgrund eines weltweiten Systems der gegenseitigen Rueckversicherung von nationalen Versicherungsgesellschaften 24 Laender in der Welt in der Lage sind, auf einer kollektiven Marktbasis Versicherungsschutz gegen nukleare RISIKEN zu gewaehrleisten. Die Kapazitaet der einzelnen Maerkte schwankt stark; sie bewegt sich zwischen einer Million US-Dollar und 85 Millionen US-Dollar. Gemeinsam koennen die vereinigten Versicherungen jedoch bis zu 160 Millionen US- Dollar an Schadensersatzanspruechen abdecken, eine Kapazitaet, die weit ueber den gesetzlichen Anforderungen der meisten Laender liegt. Lediglich in der BRD, der Schweiz und den USA gehen die gesetzlichen Anforderungn im Augenblick ueber diesen Betrag hinaus. In diesen Laendern uebernehmen jedoch entweder der Staat oder der Staat in Zusammenarbeit mit der Nuklearindustrie die Haftung fuer den Differenzbetrag aus dem von der Versicherung abgedeckten Betrag und dem Hoechstbetrag der Haftung, der vom Gesetz verlangt wird. (IAEA-Pressekommunique 84/12)

Kurznachrichten / Newlets

Fusion:

EINWEIHUNG DES JET

Die britische Koenigin Elisabeth II. weihte am 9. April 1984 im Forschungszentrum fuer Fusionsenergie in Culham bei Oxford den Joint European Torus (JET) ein. Der JET ist eine grosse Tokamakanlage zur Erforschung noch offener plasmaphysikalischer Fragestellungen. Bei der kontrollierten Kernfusion sind die Probleme der Aufheizung und Einschliessung des Plasmas, zu dem die beiden schweren Wasserstoffisotope D und T aufgeheizt und zusammengehalten werden sollen, sehr schwierig zu loesen. Die erforderlichen ausserordentlich hohen Temperaturen erlauben nicht, das Plasma in einem materiellen Behaelter zusammenzuhalten. Beim Tokamak-Konzept wird das heisse Plasma deshalb mit Hilfe von Magnetfeldern zusammengehalten. Der nun eingeweihte JET ist ein Unternehmen der Europaeischen Atomgemeinschaft(Euratom), an dem auch die Schweiz und Schweden beteiligt sind. (SVA-Bulletin (1984) 8, S.8)

Um den Zuendbereich zu erreichen, muss ein Gas aus H-Kernen bis auf 100 Mio Grad erhitzt werden. Dann erst setzen die Verschmelzungsreaktionen in grosser Zahl ein, bei denen He entsteht und die, aehnlich wie in der Sonne, einen Energieueberschuss produzieren. Bei diesen Temperaturen sollen durch das JET-Plasma, das in einem ringfoermigen Magnetfeldkaefig eingeschlossen ist, bis zu etwa 20 Sekunden lang Stroeme von bis zu 4,8 Mio Ampere fliessen - etwa zehnmal so viel wie in bisherigen Geraeten dieser Art. Der Strom wird mit Hilfe eines Transformators im Plasma induziert. Experimente dieses Typs des Plasmaeinschusses werden Tokamaks genannt. Bei den ersten "Entladungen" vom 25. Juni 1983 wurden bisher Plasmastroeme von 60 000 Ampere fuer die Dauer einer Zehntelsekunde erreicht.

Das Geraet ist etwa 12m hoch, 15m im Durchmesser und hat ein Gesamtgewicht von ca 4500 t. Zur Stromversorgung der 32 D-foermigen, knapp 6m hohen Magnetfeldspulen und des Transformators werden betraechtliche Energien eingesetzt. Neben den 600 Megawatt, die dem britischen 400-kV-Stromnetz entnommen werden koennen, wird JET zusaetzlich von zwei Schwungradgeneratoren - senkrecht rotierenden Metallzylindern von je 700 t Gewicht - mit jeweils

400 Megawatt versorgt. Diese Leistungen werden bei jeder Plasmaentladung (fuer 20s) eingesetzt. Jede der 12 t schweren Magnetfeldspulen drueckt waehrend der Entladung mit einem zusaetzlichen Gewicht von 1800 t auf die zentrale Achse. Die ringfoermige Brennkammer besteht aus 68 t. Edelstahl; der sternfoermige Eisenkern des Transformators hat ein Gewicht von 2700 t. Das JET-Plasma hat einen Ringdurchmesser von 6m, der D-foermige Plasmaquerschnitt ist etwa 4,20m hoch und 2,50 m breit.
(Naturwissenschaften 71 (1984) 94, S. 94)

Uran:
URANPREIS AUF REKORDTIEFE
Als Folge aggressiver Verkaeufe durch Uranproduzenten und Elektrizitaetsunternehmen in den letzten Monaten hat der Spotmarkt-Preisbereich fuer Natururan nach einem halbjaehrigen Abstieg im Maerz 1984 den Rekord-Tiefstand von $ 16,25-17,00/1b U_3O_8 erreicht. Als einer der Hauptgruende fuer das Ueberangebot fuer das Natururan wird die gegenwaertige wirtschaftl. Krise der Kernenergieindustrie in den USA angefuehrt.
(SVA-Bulletin (1984) 7, S. 9)

Umweltschutz:
ELEKTRONEN REINIGEN RAUCHGAS
Ein Verfahren, Schadstoffe im Rauchgas durch Elektronenbestrahlung zu entfernen, hat die amerikanische Firma EBARA International Corporation gemeinsam mit dem Japan Atomic Energy Institute und der Univ. von Tokyo entwickelt. Der Reinigungsvorgang beruht auf dem "Skrubber-Prinzip": Mit einem Waermeaustauscher oder durch Einspritzen von Wasser wird das Rauchgas zunaechst von ungefaehr 150° C auf 70 bis 90° C gekuehlt. Dann wird es mit Ammoniak versetzt und im Elektronenstrahl-Reaktor mit Elektronen beschossen. Dadurch oxidieren die Schwefel- und Stickoxide, reagieren mit dem Ammoniak und bilden Ammoniumsulfat bzw. das Mischsalz Ammoniumnitratsulfat. Begonnen hat die Entwicklung des Verfahrens mit einem kleinen Versuchsreaktor: Aus Rauchgas, das 600 bis 900 ppm SO_2 und 80 ppm NO_x enthielt, konnten 80% des Schwefeldioxids und fast 100% der Stickoxide gereinigt werden. Daraufhin baute EBARA eine Pilotanlage mit einer Durchflussrate von 1000 Nm^3/h und schliesslich wurde eine Anlage mit einer Durchflussrate von 10 000 Nm^3/h in einem japanischen Stahlwerk installiert. Waehrend eines Testmonats blieb der Wirkungsgrad konstant bei 95% fuer Schwefeldioxid und 90% fuer Stickoxide. Geplant ist jetzt das neue Programm "Process Development Units" (PDU). Es sieht eine Reinigungsanlage fuer 32000 Nm^3/h Rauchgas mit hoeherem Schwefelgehalt vor, der demjenigen von kohlebefeuerten Anlagen entspricht. Ziel dieses Projekts ist, die optimale Behandlungsart mit maximaler Wirkung und wirtschaftlicher Strahlendosis zu finden. Der Reaktor, der zunaechst 6 Monate erprobt werden soll, wird in seiner Groesse nach der Elektronenstrahlcharakteristik ausgelegt, nicht nach der Gasflussrate. So ist die Anlage fuer unterschiedliche Gasflussraten einsetzbar.
(Bild der Wissenschaft (1984) 3, S. 30)

WALDSTERBEN DURCH KERNENERGIE
Ebenso wie der Mensch unterliegt auch die Biosphaere der natuerlichen Strahlenbelastung, die sich aus den drei Anteilen kosmische Strahlung, terrestrische Strahlung (Radioaktivitaet des Bodens) und inkorporierte natuerliche radioaktive Stoffe zusammensetzt. Die Summe der drei Strahlenbelastungsanteile betraegt hierzulande beim Menschen rund 100 mrem/a. Fuer Baeume laesst sich etwa die gleiche Strahlenbelastung berechnen, wobei andere inkorporierte natuerliche Radionuklide (z.B. C-14, H-3) bedeutsam sind. Da die zusaetzliche aus Kernkraftwerken stammende Strahlenbelastung oertlich hoechstens 1%, im Mittel 0,1 bis 0,01% der natuerlichen Strahlenbelastung betraegt, ist der Beitrag der Kernkraftwerke zur Strahlenbelastung der Waelder als bedeutungslos einzuordnen. Auch die Frage, ab welcher Strahlenbelastung eine Baumschaedigung eintritt, wurde untersucht. So liegt die LD 50 je nach Baumart zwischen 500 und 8000 rem (ROENTGENSTRAHLEN). Dies entspricht uebrigens etwa dem Strahlendosisbereich fuer die Krebstherapie beim Menschen. In der Diskussion um die Waldschaeden werden als Schadstoffe neben SO_2 auch O_3 und NO_x genannt. Die Erzeugung von O_3 und NO_2 durch natuerliche und durch vom Menschen verursachte Strahlung wurde untersucht. So liegt allein die Produktionsrate von O_3 und NO_2 durch das natuerliche radioaktive Edelgas Rn und seine Folgeprodukte um mehrere Groessenordnungen ueber der Produktionsrate eines Leichtwasser-Kernkraftwerks. O_3 wird jedoch hauptsaechlich durch photochemische Prozesse (Sonnenlicht) staendig in grossen Mengen erzeugt. Allein aus der Reaktion mit NO und photochemischen Radikalen werden weltweit jaehrlich rund 10^9t O_3 gebildet. Beruecksichtigt man nur den Luftraum ueber der BRD, so wird die O_3-Produktionsrate einer 350-t/a-Wiederaufbereitungsanlage von etwa 300 t O_3/a durch die genannte photochemische Erzeugung von O_3 um etwa 4 Groessenordnungen, diejenige eines LWR-Kernkraftwerkes um 7 Groessenordnungen uebertroffen. Auch die Produktionsrate von NO_2 aus der Verbrennung fossiler Energietraeger von 3×10^6t/a ist etwa 4 Groessenordnungen groesser als die einer 350-t/a-Wiederaufbereitungsanlage und 7 Groessenordnungen groesser als die eines LWR-Kraftwerkes. Berechnete oertliche O_3- und NO_2-Konzentrationen in der Nachbarschaft kerntechnischer Anlagen liegen um mindestens 9 Groessenordnungen unter den natuerlichen O_3- bzw. NO_2-Konzentrationen in laendlichen Gebieten. Dies bedeutet, dass die direkten oder indirekten Auswirkungen von radioaktiven Emissionen kerntechnischer Anlagen fuer die Waldschaeden bedeutungslos sind. (Nach: W. Schikarski, Atomwirtschaft/Atomtechnik 29 (1984) 4, S. K+U III)

Arbeitsschutz:
BILDSCHIRM-ARBEITSPLAETZE
Zur Strahlengefaehrdung an Bildschirm-Arbeitsplaetzen teilt die Berufsgenossenschaft der Feinmechaniker und Elektrotechnik mit, dass eine solche Besorgnis, die besonders durch zwei kanadische Berichte hervorgerufen wurde, voellig unbegruendet sei. Diese Berichte seien falsch ausgelegt worden und enthielten nicht nachpruefbare Angaben. Alle messtechnischen Untersuchungen im In- und Ausland zeigten, dass der Strahlungspegel bei diesen Geraeten sehr niedrig ist. Die Geraete unterliegen in der BRD der Roentgenverordnung unter dem Kennwort "Stoerstrahler". Ihre Ortsdosisleistung darf in einem Abstand von 5 cm von der Oberflaeche 0,5 mR/h nicht ueberschreiten. Untersuchungen des KFZ Karlsruhe mit einer aufwendigen Messtechnik zeigten, dass keinerlei Stoerstrahlung festgestellt werden konnte, die gemessene Strahlung vielmehr von natuerlichen radioaktiven Stoffen aus dem Glas (K-40, Bi-214, Pb-214) stammte. Die Stoerstrahlung wird offensichtlich vollkommen abgeschirmt. (Atomwirtschaft/Atomtechnik 29 (1984) 4, S. K+U III)

Tieftemperaturphysik:
SUPRALEITER RHODIUM
Die Tieftemperaturgruppe des Instituts fuer Festkoerperforschung der KFA Juelich hat mit Hilfe einer Tiefsttemperaturanlage die Supraleitfaehigkeit des Elements Rh nachgewiesen. Der Effekt zeigte sich sehr nahe am absoluten Nullpunkt bei -273,15°C. Rh ist nun das Element mit dem bei weitem tiefsten bekannten Sprungpunkt. Es ist gleichzeitig das erste supraleitende Element mit sehr hoher paramagnetischer Suszeptibilitaet. Die wissenschaftliche Bedeutung der neuen Entdeckung liegt vor allem darin, dass gezeigt werden konnte, dass die Tendenz zum Magnetismus zwar die supraleitende Sprungtemperatur von Rh extrem reduziert, dass jedoch das supraleitende Verhalten selbst den

bekannten Theorien der Supraleitung gehorcht. (Atomwirtschaft/Atomtechnik 29 (1984) 4, S.168)

Ankündigungen / Announcements

N e w   J o u r n a l s :

"J. of Atmospheric Chemistry"
1(1983)1 (Editors: P.J. Crutzen and D.H. Ehhalt)
The Journal publishes original work on:
- observational, interpretative and modelling studies of composition and physico-chemical processes in the Earth's atmosphere, excl. air pollution problems of only local importance;
- the role of the atmosphere in biogeochemical cycles; chemical interaction of ocean, land surface, biosphere with the atmosphere;
- laboratory studies of the mechanics in homogeneous and heterogeneous transformation processes in the atmosphere;
- description of major advances in instrumentation developed for the measurement of atmospheric composition and chemical properties.

"J. of Inclusion Phenomena"
1(1983)1 (Editors-in-Chief: J.L. Atwood and J.E.D. Davies)
The Journal reports on original research into all aspects of the study of hostguest systems, in which the host structure may range from a synthetic macromolecule, through heterocyclic systems, to natural or synthetic minerals. The range of guest entities is similarly broad, comprising atomic, molecular, and ionic species. Naturally, the preparation and characterization of the compounds play a major role, but articles on the theoretical and thermodynamic aspects of the inclusion state are also included.

"Reviews in Chemical Engineering"
1(1983)1 (Editors: N.R. Amundson and D. Luss)
The main aim of the Journal is to develop new insights and to promote interest and research activity in chemical engineering and applied chemistry, as well as the application of new developments in these areas. The journal will publish authoritative articles of limited scope by leading chemical enineers, applied scientists and mathemdticians. Long articles which review an entire field will occasionally be solicited.

Isocommerz Mitteilungen
3/1984
Neue deuterierte Verbindungen:
- p- und o-Dimethyl-cyclohexan-d_{10}
- Mesityloxid-d_{10}
- n-Pentan-d_{12}
- n-Heptan-d_{16}

als Loesungsmittel fuer die IR- und NMR-Spektroskopie

[^{14}C]-Aminophenazon-Atemtest
Ein neues diagnostisches Praeparat zur Bewertung der Funktionstuechtigkeit der Leber

Grossflaechenpraeparat
Erweiterung des Sortiments (Reference sources) durch Co-57 und Mock-Iod-131 eine Kombination von Ba-133/Cs-137, anstelle der kurzlebigen Radionuklide Tc-99m und I-131.
Weitere Informationen bei Isocommerz GmbH Abt. Marktarbeit, DDR-7050 Leipzig, Permoserstr. 15

Veröffentlichungen / Publications

LISTE DER VEROEFFENTLICHUNGEN DES KERNFORSCHUNGSZENTRUMS KARLSRUHE AUS DEM JAHRE 1983
KfK 3725, April 1984, 124 S

INSTITUTE OF ISOTOPES OF THE HUNGARIAN ACADEMY OF SCIENCES (all in English)

Investigation of Ba-138 low spin states in the (n,n'γ) reaction and within the shell model - 29 pp., J. Diószegi et al. (IZIN-2-1984)

Pilot-scale irradiators for food preservation in Hungary - 7 pp., V. Stenger (IZIN-3-1984)

Design principles of techno- and socio-economically feasible pilot plants for food irradiation in developing countries - 36 pp., V. Stenger (IZIN-4-1984)

Food irradiation plants planned to be built in Hungary - 6 pp., V. Stenger (IZIN-5-1984)

IAEA Study Tour on Production, Control, and Utilization of Radioisotopes, including Radiopharmaceuticals, April 15 - May 10, 1984 in Hungary - 45 pp., Lectures held in the Institute of Isotopes (IZIN-6-1984)

CHEMICALLY AND BIOCHEMICALLY IMPORTANT ELEMENTS
P. Laszlo (Ed.)-NMR of newly accessible nuclei, Vol. 2 - Academic Press, 1983, 436 pp
D-NMR J.C.P. Smith
T-NMR A.L. Odell
B-11-NMR R.G. Kidd
O-17-NMR J.P. Kintzinger
Al-27-NMR J.J. Delpuech
Application of Si-29 NMR B. Coleman
Mg-25 and Ca-43 NMR K.J. Neurohr et al.
Co-59 P. Laszlo
Se-77 and Te-125 H.C.E. Mcfarlane et al.
Rh-103 B.E. Mann
Ag-109 P.M. Henrichs
Cd-113 NMR J.M. Armitage et al.
Th-NMR J.F. Hinton et al.

Veranstaltungskalender – Tagungen – Lehrgänge / Calendar – Meetings – Courses

In den Hallen der Schweizer Mustermesse in Basel finden folgende Fachmessen bzw. Konferenzen statt:

ILMAC '84
9. Int. Fachmesse f. Laboratoriums-/Verfahrenstechnik, Messtechnik und Automatik in der Chemie - 11.-14.9.1984

Anlaesslich der Eroeffnung am 11.9.1984 spricht Prof. W. Arber ueber "Genetik und Biotechnologie" und anschliessend wird Direktor A. Schaerli, Ciba-Geigy AG Basel, den neuen, mit Fr. 10000.-- dotierten Chemie-Sicherheitspreis 1983 dem ersten Preistraeger uebergeben. Vortragsreihen der Schweizerischen Gesellschaft fuer Instrumentalanalytik und Mikrochemie zum Thema "Voraussage analytisch relevanter Eigenschaften aufgrund der chemischen Struktur" (Tagungsleiter Prof. W. Simon, ETH Zuerich) und der Schweizerischen Vereinigung diplomierter Chemiker HTL zum Problemkreis "Chemie - Mensch - Computer" werden die an der Ilmac praesentierten Fachgebiete Laboratoriumstechnik allgemein - Laboratoriumstechnik-Messtechnik - Laboratoriumstechnik Isotopen- und Ra-Chemie in hervorragender Weise ergaenzen.
Auskuenfte: Sekretariat Ilmac, c/o Schweizer Mustermesse, Postfach CH-4021 Basel.

IPHARMEX 84
5. Int. Apotheker-Fachmesse - 11.-14.9.1984
Auskuenfte : Sekretariat Ipharmex, Postfach, CH-4021 Basel

SAMA '84
2. Int. Fachmesse von Spitzentechniken - 25.-29.9.1984
Fachgebiete: Automation, Oberflaechenbehandlung, Messung und Kontrolle sowie verwandte Hilfstechniken

FABRITEL '84
1. Int. Fachmesse fuer Fabrikationseinrichtungen in der Elektronik - 25.-29.9.1984

AUTOFACT '84
Konferenz mit Ausstellung ueber computergesteuerte Fabrikation - 25.-29.9.1984

Fuer die Loesung von Steuerungsproblemen hat sich seit geraumer Zeit ein neuer Terminus etabliert: Computer Integrated Manufacturing CIM. CIM ist ein Oberbegriff fuer die computergestuetzte Planung, Steuerung und Kontrolle von Produktionsprozessen.

International Symposium on MODERN TRENDS OF COLLOID SCIENCE IN CHEMISTRY AND BIOLOGY
(17.-19. Oct, INTERLAKEN/Switzerland)

Organized by the Association of Swiss Chemists - Selections from the Programm:

- On birth, life, and death of colloids - J.Th.G. Overbeek
- Structural relationship between monomeric surfactants and their aggregates - T. Kunitake
- Reactions at organized interfaces - H. Kuhn
- Catalysis of chemical reactions by colloidal metals and semiconductors - A. Henglein
- Phospholipid structure and phase behavior - G.G. Shipley
- The mechanism of membrane fusion by influenca virus - D.G. Wiley
- Experimental characterization of fluid interfaces in the presence of low and high molecular weight compounds - H. Sonntag
- Magnetic resonance methods for studying membrane and micelles. An Introduction - J. Seelig
- Application of neutron and X-ray scattering to the study of biological membrane structure - D.M. Engelman

Ausgewählte Beiträge / Recently published papers

ATOMNAJA ENERGIJA 57(1984)2 (all in Russian)

Parameters of Deuterium Ions Trapping into Molybdenum - A.A. Pisarev et al.

Using Miompha and Miomphosphonates for Determining the Radionuclide Content in Water Wastes - N.E. Tsvetayeva et al.

On a Method of Calculating Membrane Cascades for Multicomponent Gaseous Separation - E.B. Gruzdev

Measurement of the Range of Recoil Nuclei of Heavy Actinoids Formed in Multinucleon Transfer Reactions Produced by Ne-22 Ions - A.G. Demin

Low-Background Underground Laboratory for Radiogeochemical Measurements - Yu.A. Surkov

The Effect of Reactor Irradiation on the Microstructure of Microfuel Pyrocarbon Coatings - I.S. Alekseyeva

Express Routine Determination of Uranium in Geological Samples by Delayed Neutron Activation - V.R. Burmistrov

Isotopic Composition of Cesium Accumulated in Water-Cooled Water-Moderated Reactor Fuels - V.Ya. Gabeskiriya et al.

K-42 and K-43 Yields when Irradiating Potassium with Protons and Deuterons - P.P. Dmitriyev et al.

Implanting Low-Energy Hydrogen Ions in Lithium - O.G. Voronkov et al.

ATOMWIRTSCHAFT/ATOMTECHNIK 29(1984)3
Kernkraftwerke 1983 - Weltuebersicht - atw-Schnellstatistik

Fehlentwicklungen bei nachtraeglicher Erhoehung der Sicherheits- und Qualitaetsforderungen (Kernkraftwerk Grohnde) - D. Boettcher

Grundlagen der thermodynamischen Auslegung von Transportbehaelterlagern - H. Ernst u.a.

Erfuellt die Strahlenschutzpraxis ihren gesetzlichen Auftrag - D. Knoll

Planung und Handhabung des Strahlenschutzes waehrend der Revision - K. Gewehr

Optimierung durch Planung und Organisation - G. Gross u.a.

Beispiele von Leistungen eines Ingenieurunternehmens - L. Seyfferth

J. CHEM. PHYSICS 79(1983)12
The microwave spectra of CH_2DSH and CHD_2SH - Ch. F. Su et al.

Separation of O, C, and S isotopes by two-step, laser photodissociation of OCS - P.F. Zittel

Variational transition state theory calculations vor an atom-radical reaction with no saddle point: O+OH - S.N. Rai et al.

^2H solid state NMR of the butane/Graphite system - B. Boddenberg et al.

Determination of the spectrum of recoil energies for chlorine atoms generated by the $^{35}Cl(n,\gamma)^{36}Cl$ nuclear process - J. Chang et al.

IAEA-BULLETIN 26(1984)1
Nuclear power in developing countries - H.J. Laue

Technical co-operation for nuclear safety in developing countries - F.N. Flakus et al.

Drawing lessons from experience in technical co-operation - D.A. Kay

The Egyptian cancer project - C.B.G. Taylor

RCA activities in the Asian and Pacific Region - M. Kobayashi

Towards a regional co-ordinated programme in Latin America - C.R. O'Neal

North-South and South-South co-operation - P.M. Cate

Reliability engineering - T. Demler

Legal framework and regulatory provisions for nuclear activities - Ha-Vinh Phuong

Radiation protection in mining and milling of radioactive ores - J.U. Ahmed

Transport of radioactive materials by post - R.B. Pope

Food irradiation in Latin American countries - J.G. van Kooij

Spent fuel management today - V. Onufriev

Nuclear safety standards for quality assurance - N. Raisic

Safety in nuclear power plant operation - H.A. Wright

WISSENSCHAFT UND FORTSCHRITT 34(1984)4
Kernwaffenkrieg: Tod fuer die Welt - L.P. Feoktistov

Vom Benzol zum Benzen: Mehr als nur ein Nomenklatur-Problem - K. Lohs u.a.

а для координационной экстракции

$$\lg C_{MmB(o)} = \lg \overline{K}_{MmBn} + p \lg C_L + (m + n) \lg (C_{MmBn(b)} \gamma_{\pm}) \quad (21)$$

Таким образом, при бинарной экстракции тангенс угла наклона зависимости $\lg C_{MmBn}$ от $\lg C_{MmBn(b)}$ равен 1 независимо от заряда катиона и аниона соли, в то время как при координационной экстракции солей $\operatorname{tg} \alpha = 2$ и зависит от соотношения зарядов каттиона и аниона. Для иллюстрации на рис. 8 приводим данные по экстракции $MgCl_2$ солями (карбоксилатами) только для ЧАО четвертичных аммониевых оснований [2].

Для оценки экстракционной способности бинарных экстрагентов и сравнения с нейтральными экстрагентами рассмотрим, как и ранее, взаимосвязь константы бинарной экстракции соли с константами более простых процессов

$$K_{MmBn} = \frac{K^{n}_{(R_4N)_mB} \cdot K^{m}_{MAn}}{K^{mn}_{R_4NA}} \cdot K_n \quad (22)$$

где $K_{(R_4N)_mB}$, K_{R_4NA} — константы распределения соответствующих солей ЧАО, K_{MAn} — константа распределения экстрагируемой соли органической кислоты, а Kn — константа устойчивости комплекса MA_n в водной фазе (см. уравнение 24). Аналогичное выражение для координационной экстракции имеет вид:

$$K_{MmBn} = \frac{K_{MmBn} \cdot pL}{K^{p}_{L}} \cdot Kn \quad (23)$$

где K_L и $K_{MmBn \cdot pL}$ — константы распределения экстрагента и продукта экстракции.

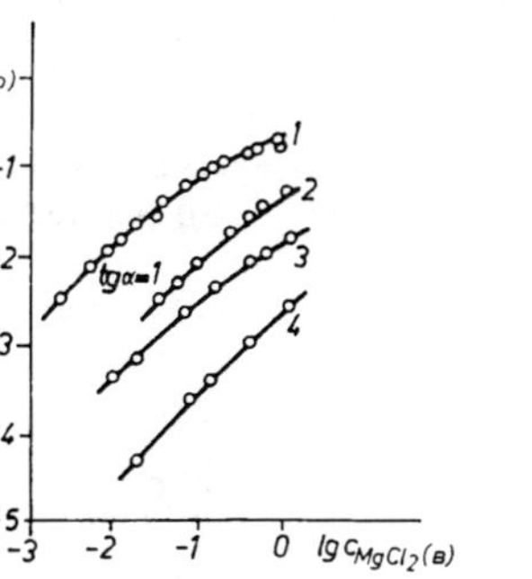

Рис. 8. Изотермы распределения хлорида магния при извлечении 0,5 М растворами карбоксилатов аминов в толуоле [3]

Исходная органическая кислота — 2-этилундеканоная; амины — 1 — триоктиламмониевая соль, 2 — 1-амино-2,2-дигексиламин, 3 — н-додецил-2-этилгексиламин, 4 — триоктиламин

Таким образом, экстракционная способность бинарных экстрагентов определяется процессом комплексообразования

$$M^{n+}_{(b)} + nA^{-}_{(b)} \rightleftarrows MAn_{(b)} \quad (24)$$

а нейтральных экстрагентов

$$mM^{n+}_{(b)} + nB^{m-}_{(b)} + pL_{(b)} \rightleftarrows MmBn \cdot pL_{(0)} \quad (25)$$

От выражения (22) можно перейти к более удобному, с точки зрения выбора экстрагентов, уравнению, связывающему константу бинарной экстракции соли с константами, описывающими исходные системы с органическими кислотами и минеральными солями ЧАО:

$$K_{MmBn} = \frac{K^{mn}_{HA} \cdot K^{m}_{M-H} \cdot K^{n}_{B-OH}}{K^{mn}_{a} \cdot K^{mn}_{A-OH}} \quad (26)$$

где K_{M-H} — константа обмена катиона металла и водорода в системе с органической кислотой.

Анализ уравнения (26) показывает, что при использовании одного и того же бинарного экстрагента и солей различных металлов с одинаковым анионом константа экстракции солей определяется константой обмена K_{M-H}, то есть порядок экстрагируемости, а следовательно и селективность зависит только от природы органической кислоты.

На рис. 9 приведены данные по бинарной экстракции нитратов различных металлов ди(2-этилгексил) дитиофосфатом тетраоктиламмония. Эктрагируемость солей падает в ряду $Cu(NO_3)_2 \gg Zn(NO_3)_2 > Ni(NO_3)_2 \gg Co(NO_3)_2 > Ca(NO_3)_2$. Таким образом порядок экстрагируемости соответствует данным по экстракции металла индивидуальными диалкилдитиофосфорными кислотами.

С другой стороны, при экстракции солей одного и того же металла с различными анионами экстрагируемость определяется значением K_{B-OH}. На рис. 10 приведены данные по бинарной экстракции солей никеля, которые показывают существенное увеличение коэффициентов распределения никеля при изменении анионов в ряду $NiSO_4 < NiCl_2 < Ni(NO_3)_2 < NiI_2$ что соответствует анионообменному экстракционному ряду для солей ЧАО.

По бинарной экстракции гидроокисей в настоящее время экспериментальных данных нет.

Для процесса

$$M^{n+}_{(b)} + n\,OH^{-}_{(b)} + R_4NA_{(0)} \rightleftarrows n\,R_4NOH_{(0)} + MAn_{(0)} \quad (27)$$

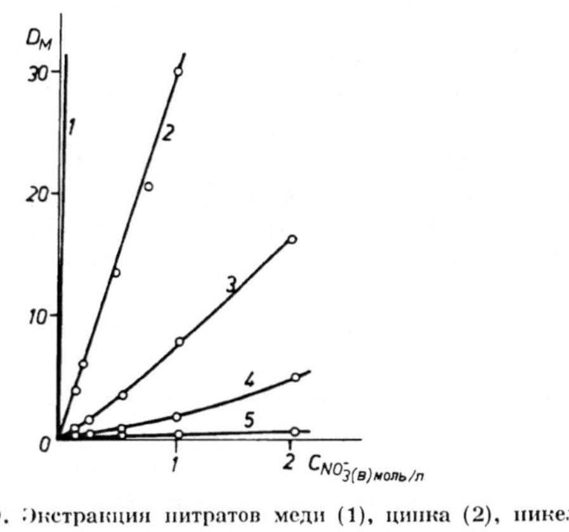

Рис. 9. Экстракция нитратов меди (1), цинка (2), никеля (3), кобальта (4) и кальция (5) 0,2 М раствором ди(2-этилгексил)-дитиофосфата тетраоктиламмония в толуоле

$C_{Cu^{2+}_{(исх)}} = 0{,}05$ моль/л; $C_{Zn^{2+}_{(исх)}} = 0{,}034$ моль/л; $C_{Ni^{2+}_{(исх)}} = 0{,}02$ мол/л; $C_{Co^{2+}_{(исх)}} = 0{,}0187$ моль/л; $C_{Ca^{2+}_{(исх)}} = 0{,}05$ моль/л

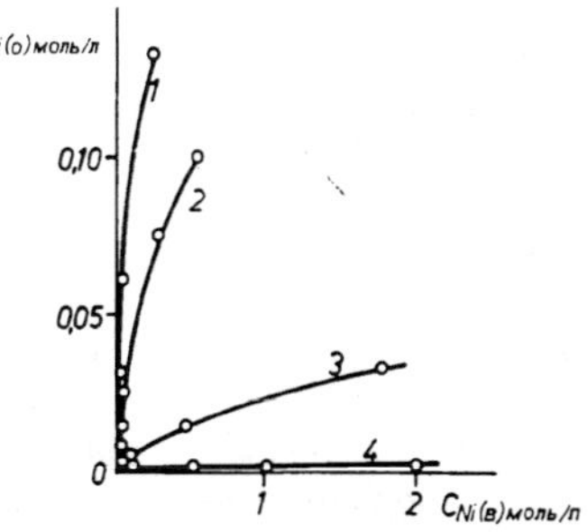

Рис. 10. Изотермы экстракции иодида (1), нитрата (2), хлорида (3) и сульфата (4) никеля 0,25 М раствором ди(2-этилгексил)-дитиофосфата тетраоктиламмония в толуоле

константа равновесия записывается

$$K_{M(OH)_n} = \frac{C^n_{R_4NOH(o)} \cdot C_{MAn(o)}}{A^n_{OH(b)} \cdot C_{M^{n+}_{(b)}} \cdot C^n_{R_4NA(o)}} \cdot \frac{\gamma^n_{R_4NOH(o)} \cdot \gamma_{MAn(o)}}{\gamma_{M^{n+}_{(b)}} \cdot \gamma^n_{R_4NA(o)}} \qquad (28)$$

Значения $K_{M(OH)_n}$ связаны с физико-химическими константами для исходных экстрагентов уравнением

$$K_{M(OH)_n} = \frac{K^n_{HA} \cdot K_{M-H}}{K^n_a \cdot K^n_{A\text{-}OH}} \qquad (29)$$

из которого видно, что экстрагируемость для одного и того же бинарного экстрагента определяется величиной константы обмена K_{M-H}.

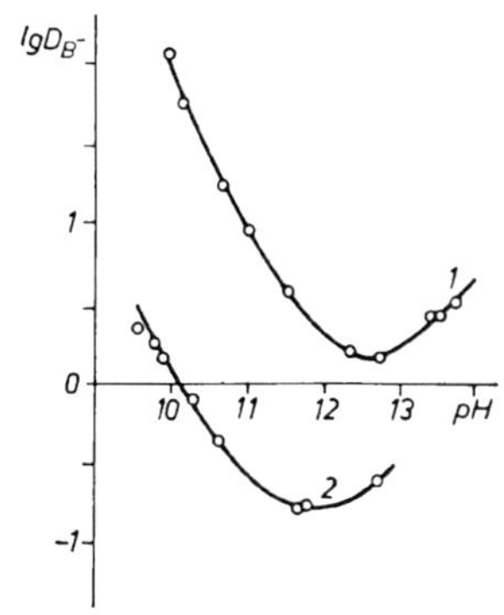

Рис. 11. Экстракция $HReO_4$ (1) и HSCN (2) 0,1 M раствором n-трет-бутилфенолята тетраоктиламмония в присутствии 0,3 M n-третбутилфенола.

$C_{ReO^-_{4(исх)}} = 0{,}005$ моль/л; $C_{SCN^-_{(исх)}} = 0{,}01$ моль/л

Следует отметить, что в системах с бинарными экстрагентами может иметь место совместная бинарная экстракция кислот и солей, а также солей и гидроокисей, что существенно усложняет экстракционные зависимости. На рис. 11 приведены данные по экстракции HSCN и $HReO_4$ n-трет-бутилфенолятом тетраоктиламмония. Как видно из рисунка, при относительно высоких pH водной фазы зависимости искривляются, что связано с бинарной экстракцией соответствующих натриевых солей в этих системах.

Кроме того, в системах с бинарными экстрагентами проявляются значительные взаимодействия между компонентами органической фазы, которые также оказывают существенное влияние на экстракционное равновесие. Так, например, при бинарной экстракции кислот наблюдается сольватация солей ЧАО органическими кислотами. Энергия взаимодействия при этом достигает 4—8 ккал/моль [5]. Аналогичное взаимодействия проявляются и при экстракции солей металлов.

Приведенные выше данные показывают, что процессы бинарной экстракции сочетают в себе ряд свойств, в целом не присущих ни одному известному ранее классу экстракционных процессов. Это существенно расширяет возможности эктракции при решении различных задач в гидрометаллургии. По-видимому, бинарные экстрагенты быстро найдут применение на практике, поскольку поведение их достаточно достоверно можно предсказать, используя обширные знания по катионо- и анионообменным экстрагентам, а получение их в ряде случаев не потребует специального синтеза.

В качестве примера, показывающего возможности бинарной экстракции, приведем экстракцию меди ди(2-этилгексил)дитиофосфатом ЧАО [7]. Этот бинарный экстрагент очень селективен к меди, что определяется более предпочтительным взаимодействием дитиофосфат-иона с катионами меди. Порядок экстрагируемости металлов для него соответствует известному ряду для ди(2-этилгексил)дитиофосфорной кислоты. Однако, если применение дитиофосфорной кислоты в качестве экстрагента меди невозможно из-за отсутствия реэкстракции металла из органической среды, то бинарный экстрагент на ее основе с успехом может быть использован для извлечения меди из растворов различного состава: реэкстракция меди легко осуществляется растворами аммиака.

Другим примером может послужить способ экстракции кобальта [8], согласно которому металл экстрагируют из аммиачных растворов алкилфенолятами ЧАО в виде комплексной кислоты с нитрозо-R-солью, а реэкстрагируют его растворами щелочей. В этом способе используется возможность получения бинарных экстрагентов с большой экстракционной способностью, что позволяет извлечь комплексную кислоту кобальта из слабощелочных аммиачных растворов. Подобная задача не может быть решена с применением известных нейтральных экстрагентов.

Таким образом, бинарная экстракция имеет ряд существенных преимуществ по сравнению с известными процессами: в ней возможно использование свойств катионообменных или анионообменных процессов, например, селективность и, в то же время, новые многочисленные пути управления процессом распределения, Кроме того в этих системах имеют место разнообразные возможности реэкстракции, как используемые для катионообменных, анионообменных и нейтральных экстрагентов, так и комбинации методов.

Процесс реэкстракции существенно облегчается ввиду образования устойчивых в органической фазе солей органических кислот и органических оснований.

Поступила в редакцию 22/VI, 1983 г.
21/XII, 1983 г.

Литература

[1] *R. R. Grinstead, T. C. Davis, S. Lynn, R. K. Charleaworth*, Ind. Eng. Chem. Prod. Res. Rev. 8 (1969) 218

[2] *T. C. Davis, R. R. Grinstead*, J. Phys. Chem. **74** (1970), 147

[3] *А. И. Холькин, В. И. Кузьмин*, Бинарная экстракция. Тезисы докладов VI Всесоюзной конференции по химии экстракции, ч. 1, Кемерово, 1981, с. 10—11

[4] *А. И. Холькин, В И. Кузьмин*, Бинарная экстракция. ЖНХ, 1982, т. 27, с. 2070—2074

[5] *В. И. Кузьмин, А. И. Холькин, В. С. Бондаренко, Н. В. Протасова, Г. Д. Холопова*, Влияние взаимодействий в органической фазе на бинарную экстракцию кислот. — Тезисы докладов VI Всесоюзной конференции по химии экстракции, ч. 1, Кемерово, 1981, с. 110

[6] *T. Sato, M. Yamamoto, N. Watanabe*. The extraction of divalent manganese, cobalt, nickel, copper, zinc, cadmium and mercury from aqueous chloride solutions by long-chain alkylammonium carboxylate. — Proc. Int. Solv. Extr. Conf., 1980, V. 2, paper 80—178

[7] *А. И. Холькин, П. К. Калиш*, Авт. свид. СССР № 596002 от 1 июня 1976

[8] *Г. А. Ягодин, В. И. Кузьмин, О. Н. Федотов, А. И. Холькин* и др., Авт. свид. СССР № 954471 от 4 мая 1982

Isotopenpraxis **20** (1984) 9, S. 345 bis 352

Трехфазная жидкостная экстракция или жидкие мембраны*

Л. Бояджиев

(Центральная лаборатория теоретических основ химической техники, Болгарская академия наук, София)[1]

Жидкие мембраны приобретают особое значение при получении и разделении веществ из сильно разбавленных растворов. Рассматриваются методы пертракционного массопереноса, для практической реализации пертракции представлены двухклеточный способ, поддерживаемые жидкие мембраны, двойноэмульсионный способ, а также пертрактор со стекающими фильмами. Приводится ряд примеров, отражающих обширную область применения, в том числе избирательное извлечение ряд металлов, минеральных и органических кислот или растворенных в воде газов. Таким же путем можно получить и биологически активные вещества, ферменты, естественные или синтетические лекарства и т. д.

Liquid membranes achieve a special importance for the yield and separation of substances from very diluted solutions. The different pertration techniques mass transfer are explained, and the two-cell methods, supported liquid membranes, double-emulsion methods as well as a pertractor with creeping films are presented. A number of examples is given, which reflect the wide range of application, as there are the selective removal of a number of metals, mineral and organic acids or water-dissolved gases. Biologically active substances, ferments, natural and synthetic medicaments and others could be separated in this way.

Keywords

membranes; separation processes; solvent extraction

Введение

Жидкие мембраны появились в инженерной химии сравнительно недавно. Благодаря своим потенциальным возможностям извлечения и разделения веществ в частности из очень разбавленных растворов, этот разделительный процесс привлек внимание многих исследователей. Появились новые наименования как: пертракция, экстракция с помощью фазы-переносителя, сочетаный транспорт, двухстадийная экстракция и т. д., несмотря на то, что речь идет об одном объекте, являющимся объединением в пространстве и времени двух знакомых процессов — экстракция и реэкстракция. Принципная схема пертракционного процесса (мы в дальнейшем будем пользоватся этим наименованием) показана на рис. 1a: перенос вещества A происходит между двумя жидкими фазами F и R, разделенными между собой третьей жидкой фазой S. Очевидно, фаза S является несмешивающейся с остальными двумя жидкостями, которые обычно близки между собой по своей природе, например водные растворы. Чаще всего встречаются трехфазные жидкостные системы типа вода-масло-вода, хотя комбинация масло-вода-масло также является предметом многочисленных исследований.

Сочетание экстракции и реэкстракции в одном процессе приводит к некоторым весьма интересным с практической точки зрения возможностям, которые нельзя осуществить при помощью классической схемы экстракции с последующей регенерацией экстрагента.

* Доклад был представлен на 3. Семинаре по Экстракции Центрального института исследований материалов, ГДР-Тюрмсдорьф, 27. 3. до 31. 3. 1983 г.

[1]) Адрес: 1040 София, Больгария

Во первых, пертракция обеспечивает максимальную разность концентраций что способствует интенсивному переносу и практически полному извлечению экстрагируемого вещества из фазы F в рамках только одной ступени контакта. Это приводит к тому, что отпадает необходимость применения противотоковых и перекрестнотоковых экстракционных схем.

Вторая очень полезная черта процесса связана с тем, что промежуточная фаза или экстрагент является кратковременным переносителем и по этой причине она не должна иметь большую емкость. К величине коэффициента распределения тоже не предявляются особые требования. Это позволяет использовать в качестве промежуточный фазы нетипичные экстрагенты или просто нейтральные жидкости с незначительной добавкой жидких и твердых активных реагентов, являющихся селективными переносителями.

Интерес к пертракционному процессу, как видно из последных обзорных публикаций ([1—6] непрерывно возрастает. Увеличивается число предлагаемых способов его осуществления, расширяется круг исследуемых объектов.

Механизм пертракционного переноса

Перенос вещества из обрабатываемой жидкости (рафинатной фазы) F к фазе-приемнику R, при участии промежуточной, нерастворимой в F и R жидкости S, осуществляется на основе одной из следующих четырех схем:

1. *Обычный физический перенос*

Обычный физический диффузионный перенос показан на рис. 1a. Примером такого типа переноса может служить разделение ароматических и алифатических углеводородов, например разделение толуола и n-гексана с использованием воды в качестве промежуточный фазы. Несмотря на то, что растворимость толуола в воде ниская — 0,51 г/л [7], то так как растворимость гексана в 50 раз ниже, толуол, если общий, объемный коэффициент массопереноса (K.a) имеет большую величину, сравнительно быстро повышает свою концентрацию в фазе R, чаще всего масло или керосин [8—11, 14—16]. При большом объеме фазы R возможно довольно понизить содержание толуола в фазе F до теоретически возможного, т. е. выравнить его концентрации в F и R.

Обычный физический транспорт несмотря на первоначальный интерес к нему и разнообразие способов осуществления, является нетипичным пертракционным процессом. Он не выгоден экономически и не представляет практического интереса.

2. *Перенос сопровождаемый химической реакцией вещества в фазе R*

Иллюстрацией этой схемы яляется извлечение фенола, аммияка, сероводорода, органических кислот и оснований и т. д. из вод при посредстве фазы S, обычного органического растворителя и их необратимого связывания в нерастворимые в фазе R соединения [20—24, 16—19]. Эта схема, показана на рис. 1б, в отличие от предыдущей, раскрывает возможности пертракционного процесса и приводит к практически полному извлечению транспортируемых веществ из исходного раствора. Например, содержание фенола в фенольных сточных водах можно уменьшить более чем на три порядка, так как образующийся в щелочной фазе R фенолят не переходит обратно. Основное преимущество этой схемы состоит в том, что в качестве ,,мембранной'' фазы можно использовать органические растворители, которые не находятся в списке экстрагентов, применяемых в знакомой экстракционной дефенолизации, экстрагенты с очень низкими (меньше единицы) коэффициентами распределения, например декан. Таким образом круг возможных экстрагентов значительно расширяется и в нем могут быть включены дешевые и экологически приемливые органические растворители. Этим же способом много веществ, образующих с добавленными в фазу R агентом нерастворимые в S продукты или осадки могут быть извлечены из раствора F почти количественно.

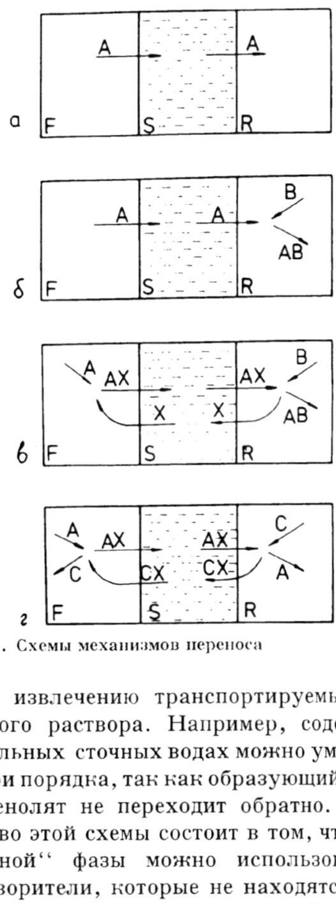

Рис. 1. Схемы механизмов переноса

3. *Промотированный перенос с химической реакцией в фазе R*

Эта схема отличается от предыдущей схемы тем, что промежуточная фаза S не является обычным, физическим растворителем переносимого вещества A, а раствором специально подобранного вещества X, которое селективно реагирует и переносит вещество A между фазами F и R. Примером могут служить случаи образования растворимого в S комплексного соединения AX, которое у фазовой границы S/R распадается, A переходит в R реагируя с B, а X возвращается, чтобы связать новые порции A у границы F/S (рис. 1в) [1, 21, 22, 24—26, 28]. В образовании переходного соединения AX могут принимать участие и другие вещества, растворенные в фазе F, как например в случае извлечения металлов из кислых вод, содержащих серную, азотную или соляную кислоту, с помощью трибутилфосфата [29]. Вместе с металлом переносятся и молекулы кислоты, которые нейтрализуются в фазе R. Возможны и другие схемы и ,,круговороты'' активного носителя и других участвующих в преносе веществ, некоторые из которых детально описаны в [5].

Эта схема демонстрирует возможность, о которой уже шла речь, использовать строго селективные, иногда специально синтезированные для переноса данного вещества, соединения в очень малых количествах, включая и твердые, нерастворимые в водных фазах F и R вещества.

4. *Обменные процессы переноса между фазами F и R*

Характерным для этой схемы является то, что транспорт вещества A из фазы F в фазу R, независимо от способа осуществления, сопровождается эквивалентным переносом вещества C из фазы R в фазу F (рис 1г). Типичным примером в этом отношении может служить ионообменная экстракция катионов металлов, которые замещаются водородными в фазе F. Место водородных ионов в фазе R занимается перенесенными в эту фазу катионами металлов [5, 16, 25—44]. Аналогически осуществяется и анионный обмен.

Здесь не будем рассматривать некоторых более комплексных пертракционных схем, как например реакции переносимого вещества в фазе S в присутствии катализатора или микроорганизмов, напр. ензимы и двухсторонная отдача продуктов реакции [47]. Интерес вызывает также наложение электрического поля, которое влияет на скорость переноса и на вид конечных продуктов [48, 49]. Последняя схема является более всего дальнейшим развитием электродиализа, чем жидкостной эктракции.

Практическая реализация пертракционного процесса

Идея практического осуществления трехфазной жидкостной экстракции наверное очень стара. Отсутствовали однако методы, гарантирующие устойчивые системы из трех жидкостей с большими межфазными поверхностями и обеспечивающие большие коэффициенты массопередачи через обоеи границы контакта. В течение последних десяти лет эта трудность была в значительной степени преодолена и было предложено несколько интересных реализаций этого процесса. Некоторые из них имеют скорее всего историческое значение и используются в основном для изучения механизма преноса в лабораторных условиях. Они не могут найти промышленного применения, так как интенсивность межфазного транспорта мала из-за небольшой удельной поверхности контакта или низких коэффициентов массопередачи. Большая часть из публикованных способов рассмотрена ниже.

1. Двухклеточный способ

Двухклеточный способ известен давно. Согласно схеме на рис. 2 он осуществляется с помощью двух сосудов с общей верхней (или нижней) частью, в которых находятся фазы F и R. Общая часть сосуда заполнена „мембранной" фазой S, которая контактирует с другими двумя фазами. Все три фазы размешиваются мешалками для повышения коэффициентов массопередачи, конечно, не допуская механического смешивания F и R [40, 50, 51]. С помощью этого способа можно

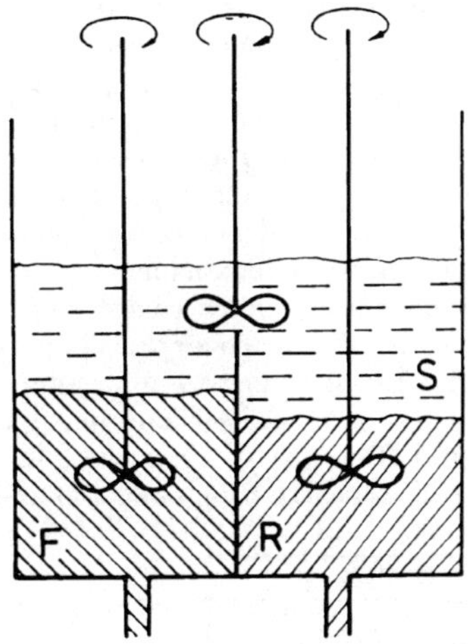

Рис 2. Схема двухячеечного способа

легко следить за ходом процесса, изучить его, но в связи с малой удельной межфазной поверхностью, обычно не превышающей 50*) $м^{-1}$. Этот способ неприемлемый для применения в производстве. Однако коэффициенты массопередачи могут быть доведены до максимальных для жидкостьной экстракции —10^{-4}m/s, если фазы F и R являются разбавленными водными растворами. Иногда исследования ведут и в четырехкамерных устройствах [52].

Модификацией этого способа является двухкамерная вертикальная колонна [4, 19, 53], схема которой показана на рис. 3. Обе камеры, разделенные фильтрирующей гидрофобной пористой перегородкой B, заполнены органической фазой-переносителем S. Водяные растворы F и R диспергируются в обоих камерах и полученные мелкие капли движутся вниз по колонне.

Под действием приложенных сбоку пульзаций, промежуточная фаза переходит из одной камеры в другую через фильтр B, не допускающий переход водяных капель. Таким образом степень извлечения определяет-

*) Рассматривается та поверхность, которая входит в меньший общий коэффициент переноса к.а и всегда относительно фазы F.

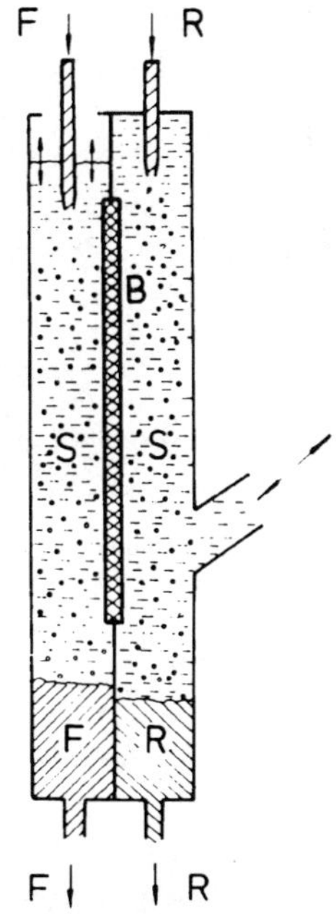

Рис. 3. Схема двухкамерной вертикальной колонны с фильтрующей перегородкой

ся не только поверхностями контакта $a_{F/S}$ и $a_{S/R}$ и коэффициентами массопереноса $K_{F/S}$ и $K_{S/R}$, а также интенсивностью межкамерного обмена фазы S, определяемой пульсационными характеристиками и объемом камер. Была разработана и конструкция с двумя фильтрами [54].

С использованием переноса бензойной кислоты из воды (F) через керосин (S) в раствор едкого натря (R), было показано, что метод обеспечивает интенсивный обмен, т. к. межфазная поверхность порядка 10^3 m^{-1}, а коэффициенты массопереноса —10^{-5} m/s. В этом методе, однако, следует иметь в виду изменение поверхностных свойств фильтрующей перегородки в связи с неизбежным присутствием поверхностно-активных веществ в системе. Надо учитывать тоже и переход мельчайших капель через фильтр.

2. Поддерживаемые жидкие мембраны

Первый вариант этого способа показан на рис. 4. Пористая твердая мембрана пропитывается фазой-переносителем S. Под действием капиллярных сил образующаяся твердо-жидкостная перегородка M разделяет находящиеся в обе ее стороны водные фазы F и R [55, 56, 41, 39, 31, 43, 57—60]. Обычно мембрана сделана из гидрофобных или гидрофобизированных материалов (полиэтилена, тефлона и др.) и имеет толщину 10—1000 μm. В последнее время используются мембраны типа милипор [58] или нуклеопор (обработанные ионами и вследствие протравления поры) [61, 62] с толщиной 10—50 μm, размером отверсий 0,02—1,0 μm и свободным сечением до 30—40%. На рис. 5 показаны наши данные о переносе меди из водного аммиачного раствора (pH_0 = 7,5) — фаза F, в разбавленный раствор серной кислоты (pH_0 = 1,5) через пористую полиэфирную мембрану типа нуклеопор, пропитанной олеиновой кислотой. Соотношение фаз F:R = 2:1.

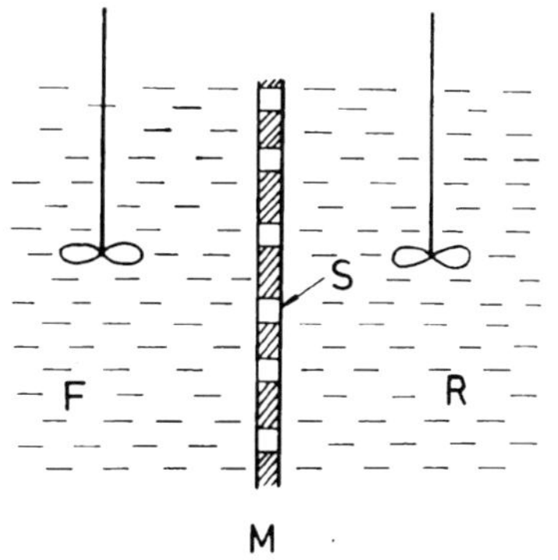

Рис. 4. Пертракция через пропитанную фазой S пористую твердую мембрану

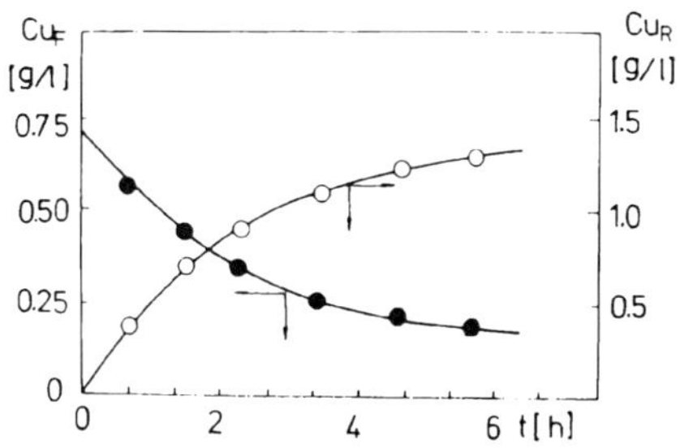

Рис. 5. Изменения концентрации меди в фазах F и R с временем для метода поддерживаемых жидких мембран

Характеристики мембраны: толщина 20 µm, рабочая площадь — 720 cm^2, диаметр пор 0,6 µm, свободное сечение 15%. Фазы F и R непрерывно циркулируют при помощи насосов. Этот метод поддерживаемых жидких мембран обеспечивает две одинаковые межфазные поверхности F/S и S/R, с площадью до 300 cm^{-1}. Так как коэффициент массопередачи в промежуточной (мембранной) фазе определяется коэффициентом молекулярной диффузии D и толщиной мембраны δ_M, очевидно его величина не может быть больше $5 \cdot 10^{-4}$ m/s. Дальнейшее уменьшение толщины мембраны ниже 10 µm не имеет смысла, так как, с одной стороны мембрана теряет свою прочность, а с другой, общее сопротивление переносного процесса уже определяется частными коэффициентами массопередачи двух прилегающих к мембране слоев воды даже при самом интенсивном перемешивании.

При реализации этого перспективного метода появляются технические трудности связанные с неидеальным начальным заполнением микропор, с неизбежной потерей фазы S из-за растворимости, с ,,пробивоом'' с мембраны в результате разных давлении по обоих сторон мембраны. Несмотря на изложенные недостатки, этот способ находит все более широкое применение например в области медицины где необходимо обрабатывать небольшие объемы растворов.

Во втором варианте этого способа используются две пористые мембраны [63]. Они необходимы для поддерживания сплошного слоя органической фазы S. Обычно мембраны из целлофана с миллимикронными отверстиями, что намного понижает общую скорость массопереноса. Повидимому осуществление такой конфигурации в промышленном масштабе затруднительно.

3. *Двойноэмульсионный способ*

Первообразом этой схемы является формирование тонких пленок промежуточной жидкой фазы вокруг движущихся в третьей жидкости капель, [64—68]. Стабилизация таких тонких пленок ($\delta < 1$ µm) возможно единственно с помощью поверхностно-активных веществ. Типичный пример в этом отношении — ,,одевание'' капель диаметром в 3—10 mm, состоящих из смеси углеводородов пленкой водного раствора сапонина. Более растворимые в воде компоненты органической смеси переходят через пленку в сплошную фазу (в данном случае R), являющуюся также органической жидкостью.

По понятным причинам эта лабораторная схема не может быть использована в практических целях, но она дала толчок развитию способов, в которых используются поверхностно-активные вещества (ПАВ).

В методе двойных эмульсий, предложенном Li [64] фаза R, по аналогии с используемыми в медицине многократными эмульсиями [69] сначала диспергируется в промежуточной фазе S до образования устойчивой микроэмулсии. Для этой цели к фазе S добалвяется эмульгатор от 1 до 5% нерастворимый в фазах F и R и имеющий гидрофильнолипофильный баланс (HLB) в пределах от 2 до 6 для систем типа вода-масло-вода и от 13 до 18 для систем масло-вода-масло [4]. Приготовление эмульсий проводится путем интензивного перемешивания с помощью высокооборотной турбины, центробежного насоса, ультразвука или другими способами. Средний размер образовавшихся микрокапель фазы обычно в границах 1—10 µm, что соответствует удельной межфазной поверхности по отношению к фазе R порядка 10^5—10^6 m^{-1}. Фазовое соотношение R/S варьирует около единицы. Полученная таким образом эмульсия диспергируется, со своей стороны, в третьей фазе F, образуя дисперсию эмульсионных глобул с диаметром 0,5—2,0 мм; как это показано схематически на рис. 6. Вторая межфазная поверхность ((R + S)/F) меньше 10^2—10^3 m^{-1}, т. к. диаметр эмульсионных глобул на два-три порядка больше.

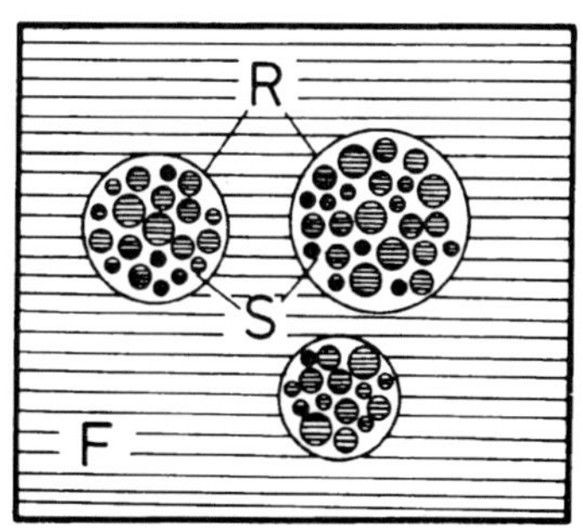

Рис. 6. Схема структуры двойной эмульсии

Необходимо подчеркнуть, что роль ПАВ в эмульсионном способе сводится к предохранению смешения однотипных фаз F и R [11], а не как считалось вначале, что они формируют некоторую ,,селективную полупроницаемую мембрану'' [9, 70]. Несмотря на то, что добавляется относительно большое количество ПАВ, эмульсия частично разрушается во время ее перемешивания в фазе F, что приводит к ухудшению степени извлечения. По этой причине отпимальное время проведения процесса, т. е. длительность перемешивания эмульсии в фазе F является строго специфическим для каждого отдельного случая в зависимости от свойств системы, фазовых соотношений, интенсивности перемешивания, конструкции аппарата и других факторов. На рис. 7 и 8 приведены два типичных примера из-

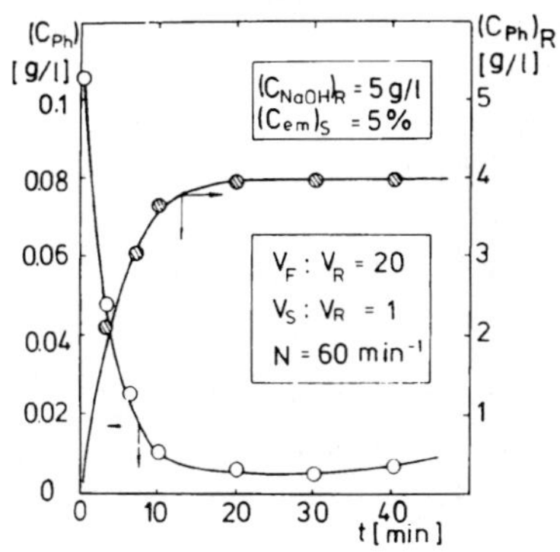

Рис 7. Кинетика извлечения фенола двойноэмульсионным способом

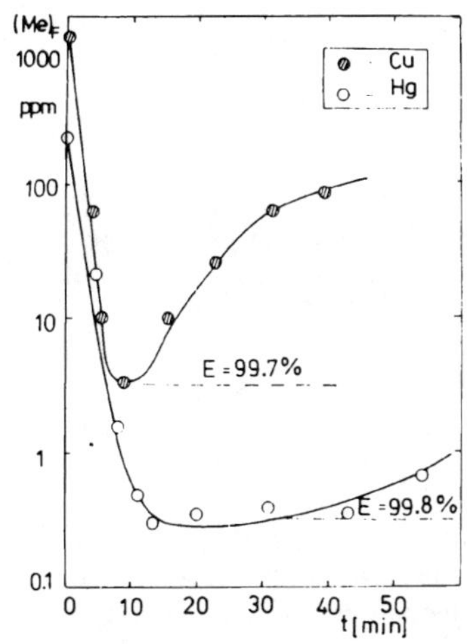

Рис. 8. Кинетика извлечения меди и ртути двойно-эмульсионным способом

влечения вещества из водных растворов с помощью эмульсионной техники. На первой из них (рис. 7) показан ход извлечения фенола, применяя как фаза S нормальные, среднекипящие парафины (C_{11}—C_{15}). Капселованный в органической жидкости —0.5% водный раствор гидроокиси натрия объемом 1/20 от фазы F играет роль приемной фазы R. Фазы R и S, в соотношении 1:1, образуют эмульсию, которая перемешивается в двадцатилитровом смесителе со скоростью 60 оборотов в минуте. Второй пример (рис. 8) показывает кинетику извлечения меди и ртути по принципу ионного обмена — схема 1г. Капселованная фаза R в этом случае — двухнормальная серная кислота, а мембранная фаза S — 10%-ный раствор высших жирных кислот в вышеупомянутых парафинах. Соотношение объемов трех фаз F:S:R = 108:1:1. При оптимальном времени проведения процесса в обоих случаях концентрации переносимых веществ в обрабатываемых водах (F) снижены более чем в 100 раз, и одновременно с этим получаются концентрированные растворы (R) этих веществ. Эмульсии, содержащие 2% ПАВ-сорбитан моноолеат разрушаются путем нагревания или электростатическим способом, после чего отделенная органическая фаза используется снова.

4. *Пертрактор со стекающими пленками*

В отличие от поддерживаемых мембран, где по крайней мере одна из фаз неподвижна, в пертракторе со стекающими пленками [71] все три жидкости находятся в движении. Это обеспечивает высокие скорости массопередачи на всех этапах переноса. Принципиальная схема этого метода показана на рис. 9. Жидкости F и R стекаются вниз под действием капиллярно-гравитационных сил по гидрофильным или гидрофилизированным поверхностям вертикально расположенных плоскостей собранных в пакет, а фаза-переноситель S циркулирует в узком пространстве между вертикальными плоскостями. В этих условиях интенсивность массопередачи на 1—2 порядка выше чем при использовании поддерживаемых жидких мембран. Возможности этого метода иллюстрируются с помощью приведенного уже примера экстракции фенола применяя как фазу S опять нормальные парафины. Следует подчеркнуть еще раз, что в системе вода-нормальные парафины коэффициент распределения фенола (m) ниже 0,1. В этом случае интенсивность циркуляции органической фазы, как показано на рис. 10, имеет решающее значение для эффективности извлечения.

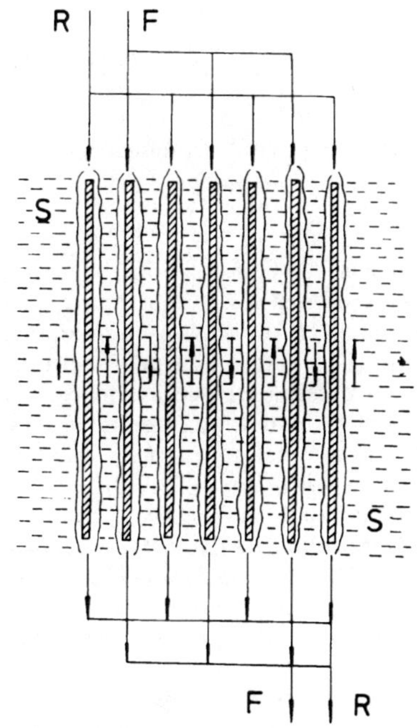

Рис. 9. Принципиальная схема пертрактора со стекающими пленками

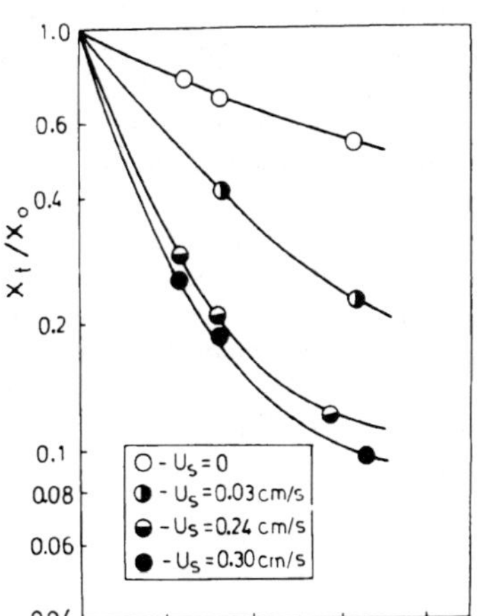

Рис.10. Кинетика извлечения фенола при помощи пертрактора со стекающими пленками. На абсцисе — времепребывание фазы F в аппарате

Моделирование пертракционного процесса

Для математического описания экстракционного процесса переноса с помощью фазы-переносителя используются два подхода:

Первый подход основан на аналогии с полимерными мембрами. Процесс переноса рассматривается как молекулярная диффузия (стационарная или нестационнарная) через плоскую или сферическую мембрану с эффективной толщиной δ_e [64, 72—77, 48, 60].

Поток массы N через эту фиктивную мембрану при обычном механизме переноса описывается уравнением [15, 17, 5]:

$$N = (D/\delta_e) \cdot a \cdot m(C_F - C_R) \tag{1}$$

где m — коэффициент распределения вещества между фазами S и F, а при переносе, осуществляемом при участие активного реагента X, растворенном в фазе S уравнением:

$$N = (D/\delta_e) \cdot a \cdot m \cdot \bar{C}_x \cdot \frac{C_F - C_R}{1 + m(C_F + C_R) - C_F C_R m^2} \tag{2}$$

где $\bar{C}_x$ — общая концентрация активного реагента X в фазе S. При $C_F \ll \bar{C}_x$ уравнение (2) сводится к:

$$N = (D/\delta_\epsilon) \cdot a \cdot m \cdot \bar{C}_x(C_V - C_R) \tag{3}$$

Этот подход, в разных модификациях используется для описания процесса переноса при почти всех использованных методах. В более комплицированных моделях стационарный процесс заменен нестационарным [72], а в случаях когда скорость процесса лимитируется химической реакцией принимается, что толщина пленки δ_e является переменной величиной, т. е. фронт реакции перемещается [37, 73].

Недостаток этого подхода заключается а априорном допущении, что общее сопротивление переносного процесса сосредоточено только в междинной фазе, так как используются величины, связанные только с этой фазой. С некоторыми исключениями эту фазу рассматривают как плоской перегородкой с постоянной толщиной без движения флуида в ней — ситуация не соответствующей действительной в используемых методах. Этот подход тоже не учитывает изменения концентрации всех веществ в объемах фаз F и S.

При втором подходе, рассматривается сочетаный перенос двух одновременно протекающих, взаимно обусловленных экстракционных процесса. Трехфазная система рассматривается как система с ограниченными фазовыми объемами и применяются известные уравнения массопереноса между двумя жидкими фазами. При этом интегральном подходе сопротивление, лимитирующее общую скорость процесса может находиться в каждом из граничных слоев или даже быть распределенным в двух или больше этапах переноса [78, 79]. Согласно этому подходу массовый поток в обычном переносном механизме определяется [78]:

$$N_i = \frac{d(V_F \cdot C_{i,F})}{dt} = K^i_{F/S} \cdot a_{F/S}(C^i_F - C^i_S/m^i_{S/F})$$

$$= -K^i_{S/R} \cdot a_{S/R}(C^i_R - C^i_S/m^i_{R/S}) \tag{4}$$

где общие коэффициенты массопереноса $K_{F/S}$ и $K_{S/R}$ у границах F/S и S/R

$$\frac{1}{K^i_{F/S}} = \frac{1}{K^i_F} + \frac{1}{m^i_{S/F} \cdot K^i_S} \tag{5}$$

и

$$\frac{1}{K^i_{S/R}} = \frac{1}{m^i_{R/S} \cdot K^i_S} + \frac{1}{K^i_R} \qquad i = 1, 2, \ldots, n. \tag{6}$$

Из уравнений (4), (5) и (6) получается

$$N = \frac{d(V_F \cdot C)^i_F}{dt}$$

$$= \frac{K^i_F \cdot K^i_S \cdot K^i_R \cdot a_{F/S} \cdot a_{S/R}\,(m^i_{F/S} \cdot C^i_F - m^i_{R/S} \cdot C^i_R)}{K^i_F \cdot a_{F/S}(K^i_R + m^i_{R/S} + m^i_{R/S} \cdot K^i_S) + K^i_R \cdot a_{R/S} \times (K'_F + m^i_{F/S} \cdot K^i_S)} \tag{7}$$

В случае двойных эмульсий, когда кроме процесса диффузии происходит и механическое разрушение эмульсии вводится также член [80, 78]:

$$N'_i = \frac{d(V_F \cdot C_{F,i})}{dt} = \alpha \cdot C^i_R \cdot V_R \tag{8}$$

где α — скорость распада эмульсии.

При допущении $K^i_R \cdot a_{R/S} \gg K'_F \cdot a_{F/S}$ уравнение (7) трансформируется:

$$N_i = \frac{d(V_F \cdot C^i_F)}{dt}$$

$$= \frac{K^i_F \cdot K^i_S \cdot a_{F/S}(m^i_{S/F} \cdot C^i_F - m^i_{R/S} \cdot C^i_R)}{K^i_F + m^i_{F/S} \cdot K^i_S} \tag{9}$$

Из уравнения (9) видно, что при $m_{S/F} \gg 1$ и $K_F \sim K_S$ общую скорость процесса определяет величина K_F, т. е. гидродинамическая обстановка в фазе F, а при $K_S \ll K_F$ и небольших $m_{S/F}$ сопротивление сосредоточено в промежуточной фазе S. В случае двойных эмульсий, принимая во внимание возможность соизмеримости двух сопротивлений, более вероятным является первый случай из-за условия $m_{S/F} \gg 1$ [4, 38, 81]. Следовательно, при вычислении необходимо применять известные зависимости для определения коэффициента массопереноса из турбулентной сплошной среды к диспергированным каплям (глобулам) или зависимости для соизмеримых диффузионных сопротивлений по обе стороны границы F/S.

Интергральный подход был в дальнейшем развит другими исследователями [38, 41, 82].

Предметом вышеизложенных рассуждений был обычный механизм переноса, в котором учитывается только диффузионная сторона процесса. При наличии медленных химических реакций, комплексообразования и др. процесс усложняется и это приводит к необходимости в учете многих дополнительных явлений. Не исключена также возможность в связи с полидисперсностью микрокапель фазы R, включенных в S, что в эмульсионном методе диффузионное сопротивление, процесса частично переносится и на границе R/S.

Очевидно, в каждом конкретном случае необходимо полное математическое описание происходящих процессов и на этой основе необходимо искать теоретически логические и опытно проверенные решения.

Заключение

Экстракция в трехфазных жидких системах обещает быть весьма перспективным процессом разделения имеющим ряд возможных областей применения. Перспективность процесса обуславливается как было сказано выше возможностью практически полного смещения равновесия в желаемую сторону при минимальных объемах междинной фазы и фазы приемника, с использованием в качестве фазы—переносителя нетипичные экстрагенты или разбавленных растворов селективных добавок. Например, при обработке водных растовор с целью извлечения из них определенных веществ могут в качестве промежуточной фазы быть использованы нейтральные масла, алканы, и т. д. с добавлением лишь нескольких процентов, а иногда и меньше одного процента, активного переносителя, который при обычной экстракции является экстрагентом. В качестве активных переносителей могут быть использованы и твердые вещества, практически нерастворимые в воде, но растворимые в фазе S. В связи с их очень нискими потерями можно применят сравнительно дорогие, высокоселективные вещества.

Специфичный, кажущийся ,,перенос вещества против градиента концентрации'' позволяет получение высококонцентрированные экономически выгодные для переработки растворы извлеченных разных ценных веществ.

К перспективным областям применения относятся:

Селективное извлечение металлов, например меди, молибдена, цинка, ванадия, вольфрама, урана, золота, ртути, рения, хрома, кадмия, никеля, лантанидов и т. д.

Этим же сопособом могут быть извлечены и концентрированы в воде соответствующих соединений фенолы, азотная, уксусная, лимонная и др. кислоты, органические основания, аммиак, сереводород и другие газы, растворенные в воде.

Особый интерес представляет селективное извлечение из культуральных и других жидкостей разных биологически активных продуктов, ферментов, ряд природных и синтетических лекарственных веществ и их концентрирование путем подбора подходящей промежуточной фазы.

Интенсивные исследования направленные на создание новых способов практического осуществления процесса а также на отыскание новых объектов применения, показывают, что в недалеком будущем пертракция станет предпочитаемым процессом разделения.

Поступила в редакцию 22/VI, 1983 г.
21/XII, 1983 г.

Литература

[1] *T. Kitagawa, Y. Nishikawa, J. Frankenfeld, N. N. Li*, Envir. Sci. and Techn. **11** (1977) 602
[2] *П. А. Попов, А. С. Грищенко, Ю. П. Панов*, Нефтопре. и нефтохимия **6** (1977) 45
[3] *W. Halwachs, K. Schügerl*, Chem. Ing. Techn. **50** (1978) 767
[4] *L. Boyadzhiev, G. Kyuchoukov*, J. Membr. Sci. **6** (1980) 107
[5] *R. Marr, A. Kopp*, Chem. Ing. Techn. **52** (1980) 399
[6] *L. Boyadzhiev*, Proceed. Intern. Congr. CHISA'81, Prague 1981
[7] *C. McAuliffe*, J. Phys. Chem. **70** (1966) 1267
[8] *N. N. Li*, Ind. Eng. Chem. (Proc. Des. Dev.) **10** (1971) 215
[9] *N. N. Li, A. L. Shrier*, in: Recent Develop. in Sep. Sci. **1** (1972) CRC. Press. Boca Raton, USA, p. 163
[10] *N. D. Shah, T. C. Owens*, Ind. Eng. Chem. (Proc. Des. Dev.) **11** (1972) 58
[11] *J. Casamatta, L. Boyadzhiev, H. Angelino*, Chem. Eng. Sci. **29** (1974) 2004
[12] *H. Shindo*, Nature **270** (1977) 5634
[13] *D. Woermann*, J. Membr. Sci. **7** (1980) 127
[14] *G. Casamatta, D. Bouchez, H. Angelino*, Chem. Eng. Sci. **33** (1978) 145
[15] *S. Schlosser, E. Kossaczky*, Proceed. PERMEA'75, Modra-Harmonia 1975
[16] *I. Strzelbcki, W. Charewicz*, Wiadom. Chem. **31** (1977) 639
[17] *R. P. Cahn, N. N. Li*, Sep. Sci. **9** (1974) 505
[18] *J. Stelmaszek*, Inz. Chemiczna **9** (1979) 211
[19] *L. Boyadzhiev, G. Kyuchoukov*, Proceed. Int. Congr. CHISA'78, Prague 1978
[20] *Ж. Казаматта, А. Анжелино, Л. Бояджиев*, TOXT **11** (1977) 825
[21] *P. Alessi, B. Canepa, I. Kikic*, Canad. J. Chem. Eng. **57** (1979) 54
[22] *S. Schlosser, E. Kossaczky*, Proceed. Int. Congr. CHISA'75, Prague 1975
[23] *S. Schlosser, E. Kossaczky*, Proceed. Int. Congr. CHISA'78, Prague 1978
[24] *R. Rautenbach, R. Albrecht, O. Machhammer*, Maschinenmarkt **78** (1981) 1910
[25] *D. K. Schiffer, E. L. Cussler, D. F. Evans, A. M. Hochhauser*, Nature **250** (1974) 484
[26] *L. Boyadzhiev, E. Besenshek*, C.r. Acad. Bulg. Sci. **29** (1976) 559
[27] *F. Caracciolo, D. F. Evans, E. L. Cussler*, AIChEJ **21** (1975) 160
[28] *J. W. Frankenfeld, N. N. Li*, Sep. Sci. **3** (1977) 285
[29] *D. I. Lunt, O. Naden*, Proceed. Int. Conf. HYDROMETALLURGY'81, Manchester 1981
[30] *R. P. Cahn, J. W. Frankenfeld, N. N. Li, D. Naden, K. N. Subramanian*, in: Recent Develop. in Sep. Sci. **6** (1981), CRC Press, B. R., USA, p. 18
[31] *R. W. Baker, M. E. Tuttle, D. J. Kelly, H. K. Lonsdale*, J. Membr. Sci. **2** (1977) 213
[32] *D. F. Evans*, Colloid Interface Sci. **5** (1978) 119
[33] *J. P. Behr*, J. Am. Chem. Soc. **95** (1973) 6108
[34] *E. M. Choy, D. F. Evans, E. L. Chussler*, J. Ann. Chem. Soc. **96** (1974) 7085
[35] *T. P. Martin, G. A. Davies*, Hydrometallurgy **2** (1976/7) 315
[36] *S. Schlosser, L. Boyadzhiev, E. Besenshek, E. Kossaczky*, Proceed. of CHISA'79, Mar. Lazne 1979
[37] *A. Kopp, R. Marr, F. Moser*, Proceed. of CHISA, 78 'Prague 1978
[38] *K. Kondo, K. Kita, I. Koida, J. Irie, F. Nakashio*, J. Chem. Eng. Jap. **12** (1979) 203
[39] *E. L. Cussler*, AIChEJ, **17** (1971) 1300
[40] *W. Völkel, W. Halwachs, K. Schügerl*, J. Membr. Sci. **6** (1980) 19
[41] *F. Nakashio, K. Kondo*, Sep. Sci. and Techn. **15** (1980) 1171
[42] *H. Oaki, M. Ishida, T. Shirai*, J. Chem. Eng. Jap. **13** (1980) 251
[43] *T. Kataoka, T. Nishiki, Y. Tamura, K. Ueyama*, J. Chem. Eng. Jap. **13** (1980) 35
[44] *L. Boyadzhiev, E. Bezenshek*, Chem. Ing. Techn. **54** (1982) 506
[45] *R. Marr, H. I. Bart, A. Banvier*, Chem. Ing. Techn. **53** (1981) 208
[46] *И. П. Золотарев, Ю. М. Попков, С. Ф. Тимашев*, Всесоюз. конф. по экстр., Рига 1982
[47] *I. W. Frankenfeld, N. N. Li*, Treatise on Anal. Chem. **5** (1982) 251
[48] *А. П. Попов, С. К. Тимофеева*, Всесоюз. конф. по экстракции, Рига 1982
[49] *Е. В. Засорин, В. П. Голубев, К. М. Доюмаев*, Всесоюзн. конф. по экстр. Рига 1982
[50] *K. Kondo, S. Takashi, T. Zsuneyuki, F. Nakashi*, J. Chem. Eng. Jap. **11** (1978) 193
[51] *H. L. Rosano, J. H. Schulman, P. Duby*, J. Phys. Chem. **65** (1961) 1704
[52] *К. Б. Яцимирский, Г. Г. Талакова*, Всесоюз. конф. по экстр., Рига 1982
[53] *P. Nevokar, J. Chernohlavek*, Proceed. CHISA '79, Mar. Lazne 1979
[54] *M. Mitrovic*, Proceed 3CEC, Graz 1982
[55] *J. Sandblom*, J. Phys. Chem. **73** (1969) 257
[56] *E. Pfeferkorn, R. J. Varoki*, Colloid Interface Sci. **52** (1975) 89
[57] *M. Igawa, M. Seno, H. Takashi, T. Yamabe*, J. Membr. Sci. **2** (1977) 263
[58] *L. Kyung-Hee, D. F. Evans, E. L. Cussler*, AIChEJ **24** (1978) 860
[59] *T. Kataoka, T. Nishiki, K. Ueyama*, Bull. Chem. Soc. Japan **55** (1982) 1306
[60] *P. R. Danesi, E. P. Horwitz, P. Rickezt*, Separ. Sci. Techn. **17** (1982) 1183
[61] *K. Spurny*, Chem. Techn. **6** (1977) 69
[62] *Г. А. Ягодин, С. Ю. Ивахно, А. В. Левкин, С. А. Соловьев*, Всесоюз. конф. по экстр., Рига 1982
[63] *V. Golubev, B. Purin, V. Chibizov*, Proceed. Intern. Congr. CHISA '81 Prague 1981
[64] *N. N. Li*, AIChEJ **17** (1971) 459
[65] *R. D. Steele, J. E. Hallagan*, Sep. Sci. **9** (1974) 299
[66] *J. Stelmaszek, B. Borkowska*, Inz. Chem. **4** (1975) 869
[67] *S. Schlosser, E. Kossaczky*, Proceed. Int. Congress CHISA, Prague 1975
[68] *W. Halwachs, E. Flaschel, K. Schügerl*, J. Membr. Sci. **6** (1980) 33

[69] *W. J. Herbert*, Lancet **1** (1965) 771
[70] *P. Alessi, I. Kikic*, Riv. Combust. **27** (1973) 400
[71] *L. Boyadzhiev, Z. Lazarova, E. Besenshek*, Proceed. IV Intern. Summer School on Chem. Eng., Varna 1982
[72] *E. S. Matulevicius, N. N. Li*, Separ. Purif. Methods **4** (1975) 73
[73] *W. S. Ho, T. A. Hatton, E. N. Lighfoot, N. N. Li*, Proceed. 2nd Congr. Chem. Eng., Montreal 1981
[74] *R. Noble*, Proceed. Intern. Chem. Eng. Congr., Montreal 1981
[75] *J. Casamatta, C. Chavarie, H. Angelino*, AIChEJ **24** (1978) 945
[76] *R. Marr, H. J. Bart, A. Bonvier*, Ger. Chem. Eng. **4** (1981) 209
[77] *L. Gladek, J. Stelmaszek, H. Szust*, Proceed. Inter. Mediterranean Congress Chem. Eng., Barcelona 1978
[78] *L. Boyadzhiev, T. Sapundzhiev, E. Besenshek*, Sep. Sci. **12** (1977) 541
[79] *S. Schlosser, E. Kossaczky, V. Gruskova*, Procces. Symp., "Modelling of Separ. Processes", Chlum. u Trebone 1980
[80] *G. Casamatta, L. Boyadzhiev, H. Angelino*, Proceed. Intern. Congress CHISA, Prague 1975
[81] *H. Savistovski, S. Ortiz*, Proceed. Conf. "HYDROMETALLURGY '81", Manchester 1981
[82] *V. J. Kremesec*, Separ. Purif. Methods **10** (1981) 117

Isotopenpraxis **20** (1984) 9, S. 352 bis 359

Серусодержащие экстрагенты для разделения благородных и цветных металлов*

В. Г. Торгов

(Институт неорганической химии СО АН СССР, г. Новосибирск)[1])

Из числа серусодержащых экстрагентов для разделения благородных и цветных металлов преимущественно употребляемыми являются органические сульфиды и сульфоксиды, полученные из нефтевых концентратов, и их окисные продукты. Рассматриваются основные закономерности экстракции благородных и цветных металлов посредством органических сульфидов и сульфоксидов, приведены экстракционные постаянные для экстракции нитрата серебра и хлорида трехвалентного золота посредством сульфидов. Конкретно разботаны и представлены в схемах способы отделения благородных и цветных металлов в гидрометаллургии благородных металлов.

Among the sulfur-containing extractants organic sulfides and sulfoxides are preferredly applicable for separating noble and non-ferrous metals, above all the available natural sulfur-organic compounds, being extracted from mineral oil concentrates, and their oxidation products are used. The fundamental principles of the extraction of noble and non-ferrous metals by use of organic sulfides and sulfoxides are treated; constants for the extraction of silver nitrate and gold (III) chloride by use of sulfides are given. Methods for the separation of noble metals from non-ferrous metals and for the separation of noble metals have been elaborated and summarized schematically.

Keywords

ion exchange; organic solvents; organic sulfur compounds; solvent extraction; transition elements

Сероорганические соединения, содержащие тиольную группу, экстрагируют внутрикомплексные соединения многих благородных и цветных металлов. Более селективны нейтральные серусодержащие экстрагенты различных классов [1]. Среди них наибольшей избирательностью по отношению к благородным металлам выделяются пока еще недостаточно изученные органические сульфиды и их производные [2]. Органические сульфиды сами по себе являются представителями генетически связанного ряда соединений

$$RSH \rightarrow R_2S \begin{cases} \nearrow R_3S^+A^- \\ \searrow R_2SO \rightarrow R_2SO_2 \end{cases}$$

Строение этих соединений и донорная способность атома серы в них определяют принципы экстракции металлов и её избирательность.

Меркаптаны проявляют свойства катионообменных [3], а соли сульфониевых оснований [4] — анионообменных экстрагентов. Эти соединения, наряду с некоторыми благородными металлами, извлекают также висмут, селен, теллур, железо, медь, кобальт, никель и др. В ряду сульфид-сульфоксид-сульфон понижается донорная способность атома серы, вследствие чего сульфоны как серусодержащие экстрагенты интереса не представляют [5]. Таким образом, из рассмотренного ряда соединений заслуживают внимания органические сульфиды и сульфоксиды особенно доступные природные экстрагенты на основе выделенных из нефтей концентратов сероорганических соединений и продуктов их окисления.

В настоящем сообщенни обобщены литературные данные и результаты автора по экстракции металлов сульфидами и сульфоксидами, на основании которых разработаны конкретные способы отделения благородных металлов от цветных и разделения благородных металлов.

Отметим лишь самые общие закономерности экстракции благородных и цветных металлов органическими сульфидами (ОС) и сульфоксидами.

Эктракция металлов органическими сульфидами. Трудно протонирующиеся ОС ($pK_a = -6{,}7$) извлекают в органическую фазу комплексные неэлектролиты наиболее халькофильных элементов группй Б — золото и палладий (независимо от природы минеральной кислоты), серебро и ртуть (HNO_3), из сульфатно-хлоридных сред с высоким содержанием H_2SO_4 экстрагируются также платина (IV), таллий(III), медь(I). Коэффициенты распределения (D) других платиновых металлов при обычных температурах и в отсутствии ускоряющих добавок в связи с кинетической инертностью ацидокомплексов в реакциях внутрисферного замещения имеют значения 10^{-2}. Еще меньшими зна-

* Доклад был представлен на 3. Семинаре по экстракции Центрального института исследований материалов, ГДР-Тюрмсдорьф, 27. 3. – 31. 3. 1983 г.

[1]) Адрес: 630 090 Nowosibirsk 90

чениями D характеризуется экстракция щелочных, щелочноземельных, цветных и редких металлов [2, 5, 6]. Для ускорения экстракции платины и платиновых металлов-спутников предложено вводить в ОС восстановители (хлориды меди(I), ртути(II), олова(II) [7]), обрабатывать водные растворы окисью углерода [8], повышать температуру [9]. Сопоставление с литературными данными [1] показывает, что ОС более избирательны для благородных металлов по сравнению с другими серусодержащими экстрагентами, в которых ответственный за экстракцию атом серы входит в состав $P = S$ и $C = S$ групп. Препятствующие координационной экстракции процессы образования ацидокомплексов золота, палладия, ртути, таллия и др. обуславливают снижение D с увеличением кислотности водной фазы. В сопоставимых условиях D большинства металлов убывают при переходе от нитратных сред к хлоридным и сульфатным, т. е. изменяются антибатно с энергией гидратации входящего в состав экстрагируемого комплекса аниона. В процессах, требующих высокой емкостью органической фазы, ОС можно использовать без разбавления. Предельные отношения металл : экстрагент (L) соответствуют образованию наиболее распространенного типа координационных соединений $[AuLCl_3]$, $[PdL_2A_2]$, $[TlL_2Cl_3]$, $[AgLNO_3]$. При разбавлении ОС хлороформом, бензолом, CCl_4, гептаном экстракция убывает в перечисленном ряду разбавителей. Во многих случаях, особенно в среде хлороформа, наблюдается восстановление металлов до более низких степеней окисления. При этом золото(I) экстрагируется более эффективно, нежели золото(III) [10]. Процесс диспропорционирования Au(I) с выделением металла протекает весьма медленно, вследствие чего не возникает каких-либо осложнений при практическом проведении экстракции.

Эффективный заряд на атоме серы, характеризующий его донорную способность [5], практически не зависит от длины радикала в R_2S. Различное влияние длины радикала на экстракцию металлов (D золота, палладия и платины снижаются с удлинением R, серебра — не зависят от длины радикала, а сулемы — увеличиваются) обусловлено стерической совместимостью сульфида и аниона в ближайшем окружении металла [6]. Хотя состояние атома серы в предельных гетероциклах (тиофаны) существенно не отличается от такового в R_2S. циклические сульфиды обладают несколько более высокой экстракционной способностью, нежели сульфиды с открытой цепью. Отрицательный индукционный эффект фенильных заместителей приводит к снижению D золота, серебра и палладия в ряду диоктилсульфид > октилфенилсульфид > дифенилсульфид (ДФС) на 3—4 порядка. ДФС по показателям экстракции лишь немногим превосходит ароматические гетероциклы (тиофен, бензотиофен и др.), донорные свойства атома серы в которых понижены, как следует из SK_β — спектров, за счет $p-\pi$ сопряжения [5]. При экстракции дитиоэфирами выполняется установленная Л. А. Чугаевым закономерность о наиболее высокой устойчивости комплексов с пятичленным циклом [2]. Макроциклические тиоэфиры мало пригодны для практического использования [1].

Рассмотренные выше серусодержащие соединения в известной степени моделируют основные классы сероорганических соединений нефтей. Установленные для индивидуальных ОС закономерности экстракции металлов полностью сохраняются при переходе к выделенным из сернистых нефтей концентратам сульфидов (сульфиды нефти — СН) как по набору экстрагирующихся металлов, так и по влиянию различных факторов на их извлечение (концентрация и природа минеральной кислоты, природа разбавителя) [2, 6, 12]. Способ выделения СН экстракцией серной кислотой [13] обеспечивает очистку от азото- и кислородосодержащих соединений нефтей. Поэтому СН в отличие от нефтепродуктов [2] не экстрагируют железа и цветных металлов. СН в соответствии с величиной эффективного заряда на атоме серы несколько превосходят по экстракционной способности диалкилсульфиды и существенно не различаются между собой при изменении температурного интервала выкипания (190—220 °C, 220—280 °C, 280—360 °C). Для низкокипящих СН (ниже 220 °C) установлено всаливание экстрагента в концентрированные растворы нитрата серебра, приводящее к снижению емкости СН [14].

Таким образом, органические сульфиды пригодны прежде всего для отделения благородных металлов от цветных и редких, а также для разделения благородных металлов.

Экстракция благородных и цветных металлов сульфоксидами. В отличие от органических сульфидов экстракция металлов сульфоксидами характеризуется многообразием протекающих процессов. Присуствие в молекуле сульфоксидов двух донорных атомов (кислорода и серы) обуславливает экстракцию как элементов группы А, так и некоторых элементов группы Б. Кроме того, вследствие сравнительно высоких основных свойств сульфоксидов ($pK_a = -2{,}1$) возможно извлечение комплексных металлокислот обеих групп элементов.

По отношению к элементам группы А, координация к которым осуществляется посредством атома кислорода, сульфоксиды являются аналогами широко применяющегося в промышленности ТБФ, но по своей экстракционной способности почти на порядок превосходят его. Обзор данных по экстракции большого числа металлов сульфоксидами приведен в [2, 6, 12, 15—17]. Как и в случае сульфидов, экстракция металлов убывает при переходе от циклических насыщенных соединений и экстрагентов на основе продуктов окисления СН (нефтяные сульфоксиды — НСО) к соединениям с открытой цепью, алкиларилсульфоксидам и диарилпроизводным. Аналогично также влияние природы аниона минеральной кислоты. Ряд эффективности разбавителей для сульфоксидов (C_6H_6, предельные углеводороды > CCl_4 > $CHCl_3$) противоположен ряду для сульфидов. Экстракция большинства металлов в форме комплескных металлокислот из солянокислых растворов возрастает симбатно с устойчивостью однозарядных анионных комплексов. Цветные металлы, находящиеся в солянокислых растворах в виде двузарядных анионов, заметно не эктрагируются сульфоксидами.

Существенным преимуществом сульфоксидов перед ТБФ и другими кислородосодержащими экстрагентами является возможность разделения платиновых металлов за счет координации посредством атома серы. Наложение процессов координационной экстракции и экстракции ацидокомплексов определяет характер зависимости извлечения платиновых металлов от концентрации HCl. Экстракция палладия практически не зависит от кислотности водной фазы вследствие уменьшения с кислотностью вклада комплексного неэлектролита PdL_2Cl_2 и возрастания вклада $(H_3OL_3)^+PdCl_3^-$ [18]. Экстракция платины(IV), родия(III), иридия(III и IV) в форме анионных комплексов увеличивается с кислотностью. В случае родия возрастание его извлечения во времени, также как и невозможность реэкстракции родия разбавленной HCl, указывают на медленное протекание реакций внутрисферного замещения с координацией сульфоксида через атом серы [19]. Существенное различие в D платиновых металлов [18, 19] положено в основу схемы их разделения [19], включающей коллективную экстракцию палладия, платины, иридия и частично родия дигексилсульфоксидом из 6 M HCl с последующей избирательной реэкстракцией водой платины и иридия. Разделение палладия и родия также проводят на стадии реэкстракции: палладия —10% раствором диметиламина, родия — раствором нитрита натрия при 70 °C.

Золото(III) извлекается сульфоксидами из хлоридных, сульфатных, нитратных и перхлоратных растворов [5]. Экстрагируемой формой из HCl является $AuCl_4^-$. Координация через атом серы кажется наиболее вероятной при экстракции из сернокислых растворов, в которых золото находится в форме аква-иона (для ТБФ в этих условиях D невелики). Сульфоксиды по отношению к золоту менее эффективны и избирательны, нежели сульфиды. Применение их целесообразно в системах с высоким окислительным потенциалом, в которых возможно окисление сульфидов.

Хотя при переходе от сульфидов к сульфоксидам донорная способность атома серы существенно понижается (эффективный заряд возрастает от 0 до +0,5e) и вследствие этого их экстракционная способность по отношению к элементам группы Б также снижается, сульфоксиды в ряде случаев оказываются более избирательными, нежели сульфиды. В частности, удается отделять палладий от серебра и ртути (коэффициенты разделения 10^3 при практически полном извлечении палладия [18]), что невозможно при использовании других серусодержащих экстрагентов.

Стехиометрия и константы экстракции. Органические сульфиды и сульфоксиды образуют с элементами группы Б мономерные и полимерные (преимущественно димерные) комплексные соединения

$AuLCl_3$, $AuLCl$, AgL_sNO_3 (s = 1, 2) $(CuLCl)_s$ (s = 1–2, L = R_2S), ML_2Cl_2 (M = Pd(II), Pt(II)), $M_2L_2Cl_4$ (M = Pd(II), Pt(II), L = R_2S), ML_3Cl_3 (M = Ir(III), Rh(III)), RuL_2Cl_3 (L = R_2S) и др.

Для платиновых металлов в связи с отсутствием равновесных данных по межфазному распределению выводы об экстракции мономерных форм сделаны на основании сопосталения электронных спектров экстрактов и растворов индивидуальных комплексов. Количественное описание экстракции удалось провести лишь для нитрата серебра и хлоридов золота и ртути (II). На примере нитрата серебра показано, что стехиометрия и устойчивость комплексов AgL_2NO_3, $(AgL_2NO_3)_2$, $(AgNO_3)_{q+2}L_{q+3}$ в значительной степени определяется природой неводной среды. При перемене разбавителя реализуются случаи от заметной диссоциации в нитробензоле до практически полной димеризации в наименее сольватирующем гептане. Закономерности изменения констант экстракции (табл. 1—3) золота и серебра для различных экстрагентов и разбавителей рассматриваются ниже с точки зрения учета сольватации экстрагентов и комплексов разбавителями.

При координации сульфидов и сульфоксидов посредством атома серы электронная плотность делокализуется на металл, что проявляется (табл. 4) в рентгеновских K-спектрах поглощения в увеличении смещения K-края (относительно элементарной серы, ΔE, эв),

Табл. 1. Константы экстракции нитрата серебра при 25 °C [14]

Экстрагент	Разбавитель	$\lg \overline{K}_{1,2}$	$\lg K^{*)}_{\text{прис.}}$	$\lg \overline{K}_{2,4}$	$\lg K_{\text{дим.}}$
$(C_4H_9)_2S$	C_6H_6	2,64 ± 0,02	2,14 ± 0,10	6,65 ± 0,15	1,4 ± 0,2
$(C_6H_{13})_2S$	C_6H_6	2,65 ± 0,01	2,52 ± 0,02	6,62 ± 0,07	1,3 ± 0,1
	CCl_4	1,65 ± 0,02	2,10 ± 0,03	5,42 ± 0,05	2,1 ± 0,1
	n-C_7H_{16}	0,30 ± 0,20	2,32 ± 0,05	5,10 ± 0,05	4,6 ± 0,5
	$CHCl_3$	4,19 ± 0,04	—	—	—
$(C_8H_{17})_2S$	C_6H_6	2,68 ± 0,01	2,54 ± 0,05	6,37 ± 0,09	1,0 ± 0,1
	CCl_4	1,83 ± 0,04	2,03 ± 0,09	4,8 ± 0,6	1,1 ± 0,6
	$CHCl_3$	4,06 ± 0,02	—	—	—
	n-C_7H_{16}	1,03 ± 0,04	2,14 ± 0,08	4,95 ± 0,15	2,9 ± 0,2

*) Константа равновесия процесса $(AgNO_3)_{q+1}L_{q+2}$ (орг.) + Ag(водн.) + NO_3^-(водн) + L(орг.) ⇄ $(AgNO_3)_{q+2}L_{q+3}$(орг.), q = 0, 1, ...

Табл. 2. Константы экстракции AgL_2NO_3 при 25 °C [14, 21, 22]

Сульфиды нефти	Разбавитель	$\lg \overline{K}_{1,2}$	Экстрагент	Разбавитель	$\lg \overline{K}_{1,2}$
190—220 °C	C_6H_6	3,19 ± 0,03	$(C_6H_5)_2S$	C_6H_6	—1,13 ± 0,02
220—280 °C	C_6H_6	3,30 ± 0,03		$CHCl_3$	—0,75 ± 0,04
	$CHCl_3$	4,60 ± 0,04	$(C_8H_{17})_2SO$	C_6H_6	—0,68 ± 0,02
280—360 °C	C_6H_6	3,29 ± 0,03		$CHCl_3$	—1,49 ± 0,01
2-амилтиофан	C_6H_6	3,00 ± 0,04		CCl_4	—1,37 ± 0,01
$(C_6H_5)_3P$	C_6H_6	10,81 ± 0,04	$(C_4H_7O)_3PO$	C_6H_6	—1,92 ± 0,05
$(C_8H_{17})(C_6H_5)S$	C_6H_6	0,53 ± 0,07	$(C_8H_{15})_2SO_2$	C_6H_6	—2,29 ± 0,02
	$CHCl_3$	1,57 ± 0,04			

Табл. 3. Константы экстракции $AuLCl_3$ при 25 °C [10]

Экстрагент	Разбавитель	HCl M	$\lg \overline{K}_{1,1}$	Экстрагент	Разбавитель	HCl M	$\lg \overline{K}_{1,1}$
ДАС	$C_6H_5CH_3$	0,5	5,01 ± 0,02	ОФС	$C_6H_5CH_3$	0,5	1,49 ± 0,05
ДАС	$C_6H_5CH_3$	6,0	4,21 ± 0,01	ДФС	$C_6H_5CH_3$	0,5	—1,00 ± 0,08
ДАС	CCl_4	0,5	3,79 ± 0,02	СН*	$C_6H_5CH_3$	0,5	4,87 ± 0,02
ДАС	CCl_4	6,0	3,13 ± 0,02	СН	$C_6H_5CH_3$	0,5	4,90 ± 0,01
ДАС	C_6H_6	0,5	5,19 ± 0,01	СН	$C_6H_5CH_3$	6,0	4,24 ± 0,02
ДАС	n-C_7H_{16}	0,5	3,14 ± 0,04	СН	CCl_4	0,5	3,93 ± 0,03
ДОС	$C_6H_5CH_3$	0,5	4,86 ± 0,02	СН	н-C_7H_{16}	0,5	3,53 ± 0,03

ДАС — $(C_5H_{11})_2S$, ДОС — $((C_8H_{17})_2S$, ОФС — $(C_8H_{17})(C_6H_5)S$, ДФС — $(C_6H_5)_2S$, СН* — 190—220 °C, СН — 220—280 °C.

Табл. 4. Способ координации сульфидов и сульфоксидов к металлу

Соединение [5, 11, 23]	z e	ΔE эв.	E(S2p) эв.	ν(S—O) $см^{-1}$
R_2S	—0,1	1,15	—	—
$(C_6H_5)_2S$	0,0	1,80		
$[Ag\{(C_6H_{13})_2S\}(NO_3)]$	—	1,80		
$[Ag\{C_8H_{17})_2S\}(NO_3)]$	—	1,80		
$[Ag\{(C_6H_5)_2S\}(NO_3)]$	—	2,30		
$[Au\{(C_3H_7)_2S\}Cl_3]$	—	2,80		
$[Pd\{(C_6H_{13})_2S\}_2Cl_2]$	—	2,00		
$[Pd\{(C_6H_5)_2S\}_2Cl_2$	0,2	2,40		
R_2SO	0,5	3,90	166,3	1055
$[UO_2\{(C_6H_{13})_2SO\}_2(NO_3)_2]$	0,5	3,90	166,0	970
$[Th\{(C_8H_{17})_2SO\}_3(NO_3)_4]$	0,55	4,20	166,7	958
$[Co\{(C_8H_{17})_2SO\}_2Cl_2]$	0,36	—	165,7	975
$[Pd_1\{(C_6H_{13})_2SO\}_2Cl_2]$	0,70	5,20	166,8	1130
Экстракт из 6,8 M HCl				960
$(Pd:(C_6H_{13})_2SO:Cl = 1:3:3)$	—	4,30		1130

в возрастании эффективного положительного заряда на атоме серы по K_α-спектрам (z, ед. e) и в увеличении энергии E(S2p) при переходе от лигандов к комплексам. В SK_β-спектрах комплексов с сульфидами понижается интенсивность максимума, соответствующего неподеленной паре 3p-электронов верхнего заполненного уровня несвязующего типа в R_2S. В SL-спектрах комплексов с сульфидами отсутствует максимум в коротковольновой части спектра (160 эв), в которой можно было ожидать переход 3d—2p. Это указывает на незначительный вклад дативного механизма комплексообразования, широко привлекающегося для объяснения рядов устойчивости комплексных соединений акцепторов группы Б с серусодержащими лигандами [23].

При координации через атом кислорода зарядовое состояние атома серы в комплексных соединениях элементов группы А с сульфоксидами практически не изменяется. Положение полосы SO в ИК-спектрах подтверждает результаты рентгеноспектральных исследований. Изучение экстракта палладия из кислых растворов свидетельствует о наличии в органической фазе, по крайней мере двух соединений — ионного ассоциата (с координацией через атом кислорода к протону) и комплексного неэлектролита (координация через атом серы), что объясняет независимость экстракции палладия сульфоксидами от кислотности водной фазы [5].

Сольватация экстрагента и комплекса. Интерпретацию изменений $\overline{K}_{r,s}$ (табл. 1—3) при перемене L и разбавителя удобно проводить на основании представления энергии сольватации молекулы как суммы энергий сольватации её фрагментов. Методом термодинамического цикла показано [24], что для данного металла логарифм $\overline{K}_{r,s}$ определяется электронными свойствами донора и акцептора $(\Delta G^0)_{вак.}$, сольватацией донорного атома экстрагента $(\Delta G^{S(0)}_{сольв.})$ и аниона в экстрагируемом комплексе $(\Delta G^A_{сольв.})$

$$-RT \ln \overline{K}_{r,s} = \Delta G = -\Sigma \Delta G_{гидр.} - s\, \Delta G^{S(0)}_{сольв.} + \Delta G^A_{сольв.} + \Delta G^0_{вак.} \quad (1)$$

Процессы межфазного распределения соли металла

$$M^{p+}(водн.) + pA^-(водн.) + sL(орг.) \rightleftarrows ML_sA_p(орг.) \quad (2)$$

и экстрагента

$$L(водн.) \rightleftarrows L(орг.) \quad (3)$$

характеризуются константами

$$\overline{K}_{r,s} = C_{r,s}/aC^s_L = K_{r,s}\gamma^s_L/\gamma_{r,s} \quad (4)$$

$$\overline{K}_L = C_L/C^L_{водн.} = K\gamma^L_{водн.}/\gamma_L \quad (5)$$

где $K_{r,s}$ и K_L — независящие от природы разбавителя константы; $\overline{K}_{r,s}$ и $\overline{K}_L$ — константы в данном разбавителе; $C_{r,s}$ и C_L, $\gamma_{r,s}$ и γ_L — концентрации комплекса, экстрагента и соответствующие им коэффициенты активности в органической фазе; a — активность соли металла, $C_L^{водн.}$ и $\gamma_L^{водн.}$ — концентрация и коэффициент активности L в водной фазе.

Количественный учет влияния среды сводится согласно (4) к раздельному определению γ_L и $\gamma_{r,s}$, логарифмы которых линейно связаны с энергиями сольватации экстрагента и комплекса. При известных значениях констант относительные изменения γ_L и $\gamma_{r,s}$ (стандартный разбавитель — CCl_4) рассчитаны по соотношениям

$$\gamma_L/(\gamma)_{L CCl_4} = (\bar{K}_L)_{CCl_4} n/\bar{K}_L n_{CCl_4} \quad (6)$$

$$\gamma_{r,s}/(\gamma_{r,s})_{CCl_4} = (\bar{K}_{r,s})_{CCl_4} \gamma_L^s n_{CCl_4}^{s-1}/K_{r,s}(\gamma_L^s)_{CCl_4} n^{s-1} \quad (7)$$

где n — число молей в 1 л насыщенного водой разбавителя [25].

В случае органических сульфидов изменения в энергии сольватации незначительны, в то время как для комплексов золота и серебра наблюдается существенное усиление сольватации с ростом сольватирующей способности разбавителя (E_T). В случае сульфоксидов сольватация и экстрагента и комплекса во всех разбавителях изменяются симбатно с E_T (табл. 5). Последовательность констант экстракции сульфидами ($CHCl_3 > C_6H_6 > CCl_4 > n\text{-}C_7H_{16}$) определяется преимущественно сольватацией комплекса, а при экстракции сульфоксидами (C_6H_6, $n\text{-}C_7H_{16} > CCl_4 > CHCl_3$) — экстрагента. Усиление сольватации комплекса нитрата серебра с сульфидами состава 1:2 хлороформом объясняет отсутствие полимеризации в органической фазе. Наиболее глубоко димеризация этого комплекса протекает в слабо сольватирующих средах (гептан, CCl_4) и тем сильнее, судя по константам димеризации (табл. 1), чем выше $\lg \gamma_{r,s}/(\gamma_{r,s})_{CCl_4}$. Наблюдаемая на опыте однопараметровая корреляция $\lg \bar{K}_{r,s} n^{s-1}$ — E_T (рис. 1) при эсктракции золота ($\lg \bar{K}_{1,1} = -15{,}78 + 0{,}61\ E_T$) и серебра ($\lg \bar{K}_{1,2} = -9{,}96 + 0{,}39\ E_T$) обусловлена линейной зависимостью $\lg \gamma_{r,s}/(\gamma_{r,s})_{CCl_4}$ от E_T (рис. 2).

Для слабо взаимодействующих с разбавителями сульфидов при построении корреляционных зависимостей согласно (1) (вклад $\Delta G^S_{сольв.}$ мало изменяется с увеличением реакционной способности L, а $\Delta G^A_{сольв.}$ изменяется симбатно с увеличением донорных свойств L) можно использовать электронные характеристики сульфидов (ΔE). Наблюдается хорошая корреляция для различных сульфидов, включая сульфиды нефти, с ΔE: в среде бензола для нитрата серебра $\lg \bar{K}_{1,2} = 8{,}44 - 5{,}32\ \Delta E$, в среде хлороформа $\lg \bar{K}_{1,2} = 11{,}14 - 6{,}47\ \Delta E$. Более высокое значение коэффициента при ΔE в уравнении регресии в среде $CHCl_3$ обусловлено возрастающим с увечилением донорной способности атома серы вкладом энергии сольватации комплекса, т. е. обычное допущение о постоянном вкладе энергии сольватации в реакционной серии в данном случае не выполняется.

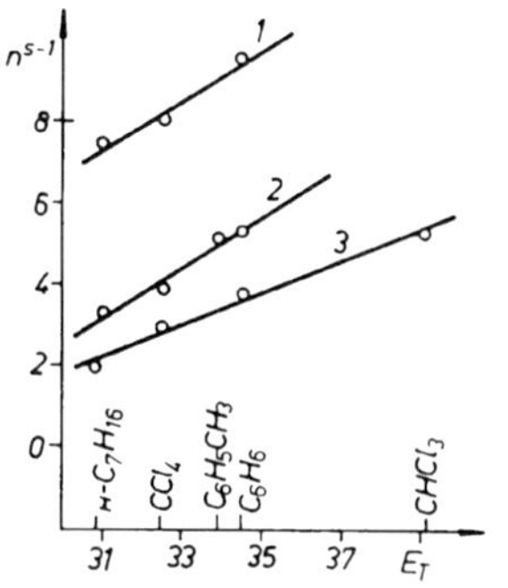

Рис. 1. Корреляции между константами экстракции серебра растворами ДОС (1, 3) и золота (из 0,5 М HCl) растворами ДАС (2) в разбавителях и E_T

1 — $\lg \bar{K}_{2,4} n^3$; 2 — $\lg \bar{K}_{1,1}$, 3 — $\lg \bar{K}_{1,2} n$

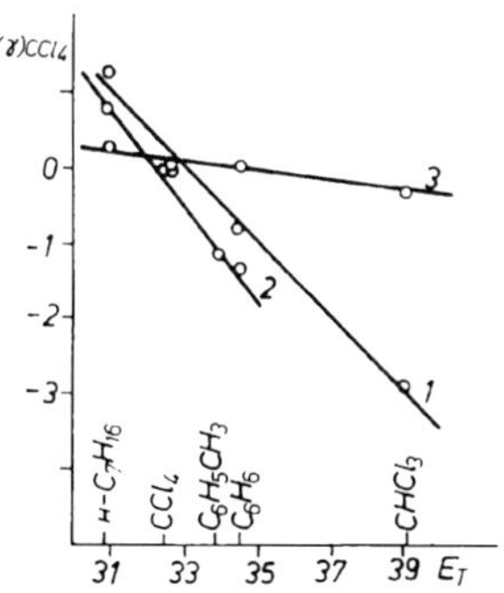

Рис. 2. Коэффициенты активности AgL_2NO_3 (1), $AuLCl_3$ (2) и L (3) в разбавителях при 25 °C (1 — ДОС, 2 — ДАС).

Для различающихся по донорному атому экстрагентов по отношению к „мягкому" серебру в среде бензола для прочных комплексов (табл. 2) сохраняется установленная для водных растворов последовательность возрастания устойчивости при переходе от кислородосодержащих лигандов-экстрагентов к серусодержащим и, наконец, к фосфоросодержащим. Для комплексов средней и низкой устойчивости принцип „жестких" и „мягких" кислот и оснований (ЖМКО) в случае диоктил- и дифенилсульфидов не выполняется. Хотя атом серы в $(C_6H_5)_2S$ по рентгеновским данным (табл. 4) является более „мягким", чем в $(C_8H_{17})_2SO$, константа экстракции для ДОСО в среде бензола выше, нежели для ДФС. Координация ДОСО к серебру осуществляется посредством атома кислорода. В среде же хлороформа наблюдается обращенность, которая

Табл. 5. Сольватация экстрагентов и комплексов [25, 26]

$\lg \gamma/(\gamma)_{CCl_4}$	$CHCl_3$	C_6H_6	CCl_4	$n\text{-}C_7H_{16}$
R_2S	−0,27	0,04	0	0,17
R_2SO	−1,60	−0,1	0	1,1
AgL_2NO_3, $L = R_2S$	−3,2	−0,9	0	1,3
$L = R_2SO$	−3,0	−1,2	0	—
$AuLCl_3$, $L = R_2S$	—	−1,4	0	0,8

Табл. 6. Способы выделения благородных металлов

Металл	Экстрагент	Применение
Палладий	R_2S	Извлечение из растоворов жидкофазного хлорирования шламов благородных металлов [29, 32]
	CH	Выделение палладия из растворов жидкофазного процесса синтеза винилацетата [36]
	R_2SO	Отделение палладия от макроколичеств азотнокислого серебра [18].
Серебро	CH	Отделение макроколичеств серебра от цветных металлов [14, 27] и получение ^{109}Cd без носителя из облученного серебра [37].
Золото	CH, HCO	Концентрирование золота и отделение макроколичеств золота от цветных металлов [2, 12].
	HCO	Извлечение золота и его отделение от цветных металлов из сернокислых тиомочевинных элюатов [38].
	R_2SO	Выделение золота без носителя из облученной платины [39].
Палладий, Платина, Иридий	HCO, R_2SO	Коллективная экстракция металлов из солянокислых растворов с последующим разделением [12, 19].
	CH	Последовательное выделение металлов за счет различия в кинетической лабильности хлорокомплексов [34].

Табл. 7. Аналитические приложения с использованием сульфидов. Методы: А—А — атомно-абсорбционный, Х—С — химикоспектральный, Н—А — нейтронно-активационный*)

Элемент, объект анализа	Метод определения	Предел обнаружения, погрешность
Золото-бедные природные и технологические продукты	А—А, распыление экстракта	0,02—0,1 мкг/мл
	А—А, ,,печь-пламя''	0,005 мкг/мл
	А—А, графитовая кювета	$2—6 \cdot 10^{-4}$ мкг/мл
	Н—А, малые навески	$1 \cdot 10^{-12}$ г (20 часов)
	Н—А, предварительное концентрирование	$1 \cdot 10^{-10}$ г (30 минут)
	Изотопное разбавление по ^{195}Au	$5 \cdot 10^{-10}$ г
Золото высокой чистоты	Х—С, 26 микропримесей	$10^{-8}—10^{-6}$%, $S_r = 0,07—0,3$
	Н—А, 28 микропримесей	$10^{-9}—10^{-6}$%, $S_r = 0,07—0,15$
Серебро-бедные продукты	А—А, распыление экстракта	0,003 мкг/мл
	Н—А, по ^{110m}Ag (0,68 мэв)	$1 \cdot 10^{-10}$ г (48 часов)
Серебро высокой чистоты	Х—С, 22 микропримесей	$10^{-8}—10^{-6}$%, $S_r = 0,07—0,25$.
	Н—А, 30 микропримесей	$10^{-9}—10^{-6}$%, $S_r = 0,07—0,15$
Палладий — бедные продукты**)	А—А, распыление экстракта	0,02—0,1 мкг/мл
	Н—А, по дочернему изотопу ^{109m}Ag (22 кэв)	$5 \cdot 10^{-10}$ г (10 часов)
Палладий высокой чистоты	Х—С, 26 микропримесей	$2 \cdot 10^{-8}—5 \cdot 10^{-6}$%, $S_r = 0,08—0,33$
	Н—А, 28 микропримесей	$10^{-9}—10^{-6}$%, $S_r = 0,07—0,15$
Палладий в продуктах на основе серебра	Х—С, экстракция сульфоксидами	$5 \cdot 10^{-7}$%

*) В скобках приведено вермя облучения, поток нейтронов $1 \cdot 10^{13}$ н/см. сек

**) Палладий определяется одновременно с золотом

обусловлена существенным увеличением сольватации сульфоксида (в отличие от сульфидов) хлороформом по сравнению с бензолом (табл. 5). Приведенный пример иллюстрирует условность принципа ЖМКО, а также ограниченность априорной оценки комплексообразования в конденсированных средах на основании электронных параметров лигандов.

Рассмотренные выше закономерности по экстракции металлов органическими сульфидами и сульфоксидами послужили основой для аргументированного определения перспективных областей применения этих экстрагентов.

Приложения экстракции металлов сульфидами и сульфоксидами. Органические сульфиды и сульфоксиды, особенно СН и НСО, являются потенциально доступными и дешевыми экстрагентами. Разработаны регламенты их технологического получения и выпущены опытные партии [13]. Сульфиды характеризуются высокой эффективностью извлечения и большой емкостью по металлу. Для извлечения золота и палладия более чем на 99% при соотношении объемов водной и органической фаз (В:О) 10:1, как показывает опыт, достаточно одной ступени экстракции, в случае серебра при В:О = 2:1 необходимо 3 ступени [27]. Это согласуется с расчетами на основе приведеных в табл. 2—3 констант. Емкость 25% (по объему) раствора СН в полиалкилбензоле по золоту составляет 140 г/л [2], 50% СН в керосине по серебру —130 г/л [27], 10% СН в изооктане по палладию —23 г/л, а 25% ДОС в „Isopar" —32 г/л [29].

Для реэкстракции золота наиболее удобен способ с использованием щелочного раствора сульфита натрия [30], сочетающий полноту реэкстракции (более 99% за одну ступень) с простотой выделения золота из реэкстракта подкислением его до pH = 1—2. Палладий реэкстрагируют 10% раствором аммиака [31], восстанавливают в экстракте водородом или отгоняют экстрагент и разбавитель [32], серебро — сульфитом натрия или проводят твердофазную реэкстракцию карбонатом натрия [14, 27]. Некоторые способы выделения и разделения благородных металлов перечислены в табл. 6.

Экстракционные технологии [32] и [29, 33] переработки растворов хлорирования шламов принципиально аналогичны (схема первоначально была опубликована в [32]). Они включают раздельное выделение золота (ТБФ [32] или дибутилкарбитолом [33]), палладия — диалкилсульфидами за одну ступень при О:В = 1:10, платины — аминами различного строения (третичными [32] или вторичными [33]). Выделенный из аммиачного реэкстракта палладозаммин содержал более 99,95% основного вещества [29]. Последовательное выделение платиновых металлов с использованием одних сульфидов введением двухлористого олова и варьированием температуры технически трудно осуществимо [34]. Схема разделения платиновых металлов экстракцией дигексилсульфоксидом [19] также достаточно сложна. Кроме того, в солянокислых растворах наблюдается химическая деструкция сульфоксидов [35]. В процессах извлечения серебра из богатых растворов СН во избежание окисления экстрагента содержание азотной кислоты не должно превышать 2,5 М [14], а реэкстракцию целесообразно проводить водой с доизвлечением серебра сульфитом натрия [27].

Применительно к анализу разработан комплекс экстракционно-инструментальных методов с использованием сульфидов, основные характеристики которых приведены в табл. 7 [40]. При А—А и Н—А определении малых (вплоть до кларковых) количеств золота, серебра и палладия избирательная экстракция этих металлов ОС, в отличие от других экстрагентов, полностью устраняет мешающее влияние сопутствующих элементов. Вследствие этого оказывается возможным без дополнительной очистки прямой А—А анализ экстракта, а в случае Н—А методов отпадает надобность в определении химического выхода. Кроме того, пределы обнаружения снижаются за счет 20—50 кратного абсолютного концентрирования. При анализе золота, серебра и палладия высокой чистоты на микропримеси в процессе удаления макроосновы не наблюдается соэкстракции микропримесей. Относительное концентрирование 10^4—10^5 (Х—С окончание) и 10^8—10^9 (Н—А окончание) получено при 1-2-х кратной экстракции растворами ОС при мольных отношениях металл: ОС = 1: 2—4. Разработанные методы широко применяются для аналитического контроля производства.

Приведенные примеры иллюстрируют большие и еще не в полной мере реализованные возможности применения сульфидов и сульфоксидов в гидрометаллургии и анализе благородных металлов. Другие области применения СН и НСО (присадки к смазочным маслам, антиокислители, флотореагенты и др.) рассмотренны в монографии [41].

Изложенный в этом докладе материал основан на совместных работах с *А. В. Николаевым*, *В. А. Михайловым*, *Л. Н. Мазаловым*, *И. Г. Юделевичем*, *Э. Н. Гильбертом*, *С. С. Шацкой*, *М. К. Дроздовой*, *Т. М. Кордой*, *В. Н. Андриевским*, *В. Д. Юматовым*, *Э. А. Гальцовой* и др.

Поступила в редакцию 22/VI, 1983 г.
21/XII, 1983 г.

Литература

[1] *Н. Г. Ванифатова, Н. В. Серякова, Ю. А. Золотов*, Экстракция металлов нейтральными серусодержащими соединениями. М., „Наука" 1980, 85 с.

[2] *В. А. Михайлов, В. Г. Торгов*, В кн. Сернистые и кислородные соединения нефтей. М., „Химия" 1971, с. 182–196

[3] *Н. Н. Артюхин, Е. А. Старцева, Л. Я. Марочкина* и др., „Радиохимия" **15** (1973) 209

[4] *E. L. T. M. Spitzer, J. Redder*, Advances in extractive metallurgy. London 1968, p. 669–677

[5] *А. В. Николаев, В. Г. Торгов, В. Н. Андриевский* и др., В кн.: Химия процессов экстракции. М., „Наука" 1972, с. 75–80

[6] *В. Г. Торгов*, Координационная экстракция металлов. Докторская диссертация. Новосибирск 1978

[7] *А. В. Николаев, Р. И. Новоселов, А. М. Еренбург, Б. И. Пещевицкий*, Авторское свидетельство 327817 – „Открытия. Изобрет. Пром. образцы. Товарные знаки" 1976, № 35, с. 153

[8] *О. С. Шелковникова, Н. К. Еременко, В. А. Михайлов*, „Изв. СО АН СССР, сер. хим. н." 1979, № 12, вып. 5, с. 97–100

[9] *Л. Н. Булганина, Г. В. Кузмичев, В. С. Колесницын* и др., X Всес. совещ. по химии, анализу и технологии благородных металлов. Тез. докл., Новосибирск 1976, част 1, с. 69

[10] *В. Г. Торгов, Т. М. Корда, И. Г. Юделевич*, „Ж. аналит. химии" 1978, т. 32, вып. 12, с. 2341–2348

[11] *Л. Н. Мазалов, В. Д. Юматов, В. В. Мурахтанов* и др., Рентгеновские спектры молекул. Новосибирск, „Наука" 1977, 312с.

[12] *В. А. Михайлов, В. Г. Торгов, А. В. Николаев*, „Изв. СО АН СССР, сер. хим. н." 1973, № 7, вып. 3, с. 3–15

[13] *М. Ф. Бондаренко, Н. К. Ляпина, Ю. Е. Никитин* и др., „Изв. СО АН СССР, сер. хим. н," 1973, № 7, вып. 3, с. 16–22

[14] *В. Г. Торгов, С. С. Шацкая, В. А. Михайлов* и др., „Изв. СО АН СССР, сер. хим. н." 1973, № 7, вып. 3, с. 79–85

[15] *M. T. Chaieb, G. Laurence*, Energie nucleaire **13** (1971) 106

[16] *Ю. Е. Никитин, Ю. И. Муринов, А. М. Розен*, „Успехи химии" 1976, т. 45, вып. 12, с. 2233–2252

[17] *V. A. Mikhailov*, CIM Spec. Vol. **21** (1979) 24

[18] *А. В. Николаев, В. Т. Торгов, В. А. Михайлов* и др., „Изв. СО АН СССР, сер. хим. н." 1967, № 14. вып. 6, с. 120–122

[19] *P. A. Lewis, D. F. S. Morris, E. L. Short*, J. Less-Common Metals **45** (1976) 194

[20] *I. Ahmad, I. Ahnad, D. F. S. Morris*, – Analyst, 102 (1977) 17

[21] *В. Г. Торгов, С. С. Шацкая, В. В. Михайлов* и др., „Изв. СО АН СССР, сер. хим. н." 1973, № 7, вып. 3, с. 70–78

[22] *В. Г. Торгов, С. С. Шацкая, Э. А. Гальцова* и др., X Всес. совещание по химии, анализу и технологии благ. металлов. Тез. докл., Новосибирск 1967, ч. 1, с. 68

[23] *Э. А. Гальцова, Л. Н. Мазалов, В. Д. Юматов* и др., „Изв. СО АН СССР, сер. хим. н." 1975, № 7, вып. 3, с. 41–45

[24] *V. G. Torgov, V. A. Mikhailov, M. K. Drozdova* et al., In: Proceedings ISEC 1974, vol. 1, p. 849–870

[25] *В. А. Михайлов, С. С. Шацкая, Д. Д. Богданова, В. Г. Торгов*, „Изв. СО АН СССР, сер. сим. н.“ 1976, № 7, вып. 3, с. 34–43

[26] *В. Г. Торгов, М. К. Дроздова, Г. А. Мардежова, З. А. Ларионова*, „Изв. СО АН СССР, сер. хим. н.“ 1980, вып. 4, с. 76–81

[27] *В. А. Михайлов, С. С. Шацкая, В. Г. Торгов, Н. И. Антипов* IX Всес. совещ. по химии, анализу и технологии благородных металлов. Тез. докл. Красноярск, 1973, с. 43

[28] *Д. М. Чижиков, Б. П. Крейнгауз, Г. М. Денисова*, „Изв. СО АН СССР, сер. хим. н.“ 1970, № 9, вып. 4, с. 120–123

[29] *I. E. Barnes, I. D. Edwards*, "Chem. Industry" **1982**, 151

[30] *А. В. Николаев, Б. И. Пещевицкий, А. М. Еренбург* и др., Авторское свидетельство № 263578. Бюлл. изобр., 1979, № 41, с. 261

[31] *Г. В. Кузмичев, В. С. Чекушин*, Авторское свидетельство № 203241. Бюлл. изобр., 1968, № 27, с. 184

[32] *В. Ф. Борбат*, В кн.: Анализ и технология благородных металлов. М., „Металлургия“ 1971, с. 381–384

[33] *R. I. Edwards*, CIM Spec. Vol. **21** (1979) 24

[34] *Э. Е. Раковский, М. И. Старожицкая*, „Ж. аналит. химии“ 1974, т. 29, № 11, с. 2094–2096

[35] *Ю. Г. Фролов, А. Г. Моргунов, В. А. Седельников, А. Ф. Меркушев*, „Изв. СО АН СССР, сер. хим. н.“ 1973, № 7, вып. 3, с. 67 до 69

[36] *М. В. Воротников, В. В. Антонова, Т. М. Кулинович*, В кн.: Совершенствование технологических процессов на предприятиях цветной металлургии Красноярского края. Красноярск 1974, с. 81

[37] *Е. С. Гуреев, Т. И. Исламов, В. С. Усаченко, Э. П. Гильберт*, „Радиохимия“ 1979, т. 21, № 3, с. 422–426

[38] *А. С. Черняк, Г. И. Смирнов, А. С. Боброва* и др., „Изв. СО АН СССР, сер. хим. н.“ 1973, № 7, вып. 3, с. 42–47

[39] *Е. С. Гуреев, Н. И. Мамадалиев*, Вторая республ. конференция молодых физиков. Тезисы докл., Ташкент, „ФАН“ 1974, с. 69

[40] *В. Г. Торгов, Э. П. Гильберт, И. Г. Юделевич, В. А. Михайлов*, Вторая Всес. конф. по методам концентрирования в аналитической химмии. Тезисы докл. М., „Наука“ 1977, с. 21–23

[41] *Я. Б. Чертков, В. Г. Спиркин*, Сернистые и кислородные соединения нефтяных дистиллятов. М., „Химия“ 1971, 307 с

Isotopenpraxis **20** (1984) 9, S. 359 bis 361

Tagungsberichte

3. Arbeitstagung „Isotope in der Natur“ vom 15. bis 18. November 1983 in Leipzig

Das Zentralinstitut für Isotopen- und Strahlenforschung (ZfI) der Akademie der Wissenschaften veranstaltete die 3. Arbeitstagung „Isotope in der Natur“. An dem Symposium beteiligten sich 82 Wissenschaftler, darunter 20 Fachkollegen aus sieben anderen Ländern (Sowjetunion – 7, VR Bulgarien – 2, VR Polen – 3, BRD – 4, Kanada – 1, Niederlande – 2, USA – 1). In sechs Plenarvorträgen wurde ein Überblick über gegenwärtig international relevante Probleme der Isotopenforschung gegeben. Die 59 Kurzvorträge und 20 Posterbeiträge gestalteten sich zu einer breit angelegten Diskussion neuester Ergebnisse auf verschiedenen Gebieten der Isotopenforschung. Der wissenschaftliche Rahmen der Arbeitstagung umfaßte

- Probleme der Verteilung stabiler Isotope und radioaktiver Nuklide in der Natur,
- theoretische und experimentelle Untersuchungen über Isotopieeffekte,
- Beiträge der Isotopengeochemie und deren Anwendung auf regionale und globale Stoffkreisläufe,
- Geochronologie und Altersbestimmung sowie
- proben- und meßtechnische Aspekte.

Das umfangreiche Vortragsprogramm wurde in Themenkomplexe, wie Isotopengeochemie, Theorie der Isotopieeffekte, Isotopenhydrologie, Radiogeochronologie und Methodik, gegliedert und weitestgehend geschlossen in Vor- und Nachmittagssitzungen behandelt. Es zeigte sich erneut, daß in der Vielfalt der wissenschaftlichen Darstellung von Ergebnissen der Isotopenforschung die Leipziger Veranstaltung sowohl in der DDR als auch im RGW-Rahmen unikal ist.

Einen wesentlichen Teil der Tagung nahm die Isotopengeochemie mit drei Plenarvorträgen und 13 Kurzvorträgen ein. Hervorzuheben sind auf diesem Gebiet die weltweit verstärkt geführten Untersuchungen zur Wechselwirkung Kruste/Mantel im Zusammenhang mit minerogenetischen Prozessen sowie die Untersuchung großregionaler bzw. globaler Kreisläufe. Im Eröffnungsvortrag hob *K. Wetzel* (ZfI) die Bedeutung der Isotopengeologie bei der Aufklärung globaler geologischer Prozesse hervor. Am Beispiel des kontinentalen Wachstums konnte am Modell der Kristallisationsdifferentiation gezeigt werden, daß aufgrund von Sauerstoff-Isotopenvariationen und Stoffverhältnissen die an Subduktionszonen abtauchende ozeanische Kruste als Quelle des überwiegenden Teils der Kontinente anzusprechen ist. Diese Schlußfolgerung wird anhand geophysikalischer, geologischer und geochemischer Überlegungen überprüft, und Kriterien zur Charakterisierung primärer Magmatite werden abgeleitet. Als Ursache für die isotopische Inhomogenität des Oberen Mantels führte *J. Hoefs* (Göttingen, BRD) in seinem Plenarvortrag die Subduktion und Isotopenfraktionierungsprozesse, wie partielles Schmelzen, fraktionierte Kristallisation, Entgasung, Verwitterung und Hochtemperaturwechselwirkung mit Seewasser, an.

Die Bearbeitung geologischer Probleme durch Kombination von Isotopen mehrerer Elemente, wie Wasserstoff, Sauerstoff, Kohlenstoff, Stickstoff, Schwefel und Strontium, unter Einbeziehung von Spurenelementgehalten hob sich als ein weiterer, auch international zu beobachtender Trend deutlich ab. Die Aktivitäten des ZfI dazu wurden durch zwei Kollektivarbeiten unterstrichen. *G. Stiehl* u. a. (ZfI) legten in einem Plenarvortrag isotopen- und elementgeochemische und geochronologische Ergebnisse über Untersuchungen an Graniten und am Nebengestein im Kontaktbereich des Ehrenfriedersdorfer Granits (Erzgebirge) dar. Im Ergebnis der komplexen Bewertung wird dieser Granitkörper als Differentiat subduzierter ozeanischer Kruste charakterisiert, wobei die Rolle des Wassers in diesem Prozeß hervorgehoben wird. Aus isotopen- und elementgeochemischen sowie thermometrischen Untersuchungen an Gas-Flüssigkeitseinschlüssen und Mineralen einer hydrothermalen Fluoritmineralisation leitete *G. Strauch* u. a. (ZfI) einen zyklischen Abscheidungsmechanismus für eine paragenetisch einheitliche Abfolge der Mineralisation ab. Die komplexe Untersuchung von Tiefseebohrungen mittels O-, H- und Sr-Isotope stellte *H. Friedrichsen* (Tübingen, BRD) eindrucksvoll zur Diskussion. Es konnte der Nachweis erbracht werden, daß Meerwasser bis in Bereiche der Moho zirkulieren kann und durch erheblichen Stofftransport Bedeutung für die Lagerstättenbildung hat. *S. Brandt* u. a. (Irkutsk, SU) unterstrichen in ihrem Vortrag die Wichtigkeit der Neodym- und Strontiumisotopenverhältnisse für die genetische Beurteilung von Alkaliultrabasiten und Mineralisationen der Seltenen Erden.

Die Möglichkeiten einer mathematischen Modellierung von Naturprozessen zeigte *G. Weise* u. a. (ZfI) für die Evolution des Schwefels und seiner Isotopenverhältnisse für den gesamten Zeitraum der Krustenentwicklung. *O. Gebhardt* u. a. (ZfI) erklären in ihrem Modell die zeitliche Änderung des $\delta^{18}O$-Wertes von Sedimenten im Diagenese/Metamorphoseprozeß. Den Langzeiteinfluß von CO_2 aus der Verbrennung fossiler Energieträger auf den natürlichen CO_2-Kreislauf bestimmten *W. Mook* u. a. (Groningen, Niederlande) aus eigenen Messungen an Proben vom Südpol und aus der Pazifikregion. Von Bedeutung für die Interpretation von Stickstoff in Gesteinen sind auch die von *H.-M. Nitzsche* u. a. (ZfI) im System N^{3-}/N_2 bei Temperaturen >1000 K experimentell gefundenen erheblichen Isotopieeffekte des Stickstoffs.

In den Vorträgen zur Theorie der Isotopieeffekte war der international erkennbare Trend auf eine Weiterentwicklung von Methoden der Quantenmechanik, der Streutheorie und der statistischen Physik und auf eine Anwendung geeigneter Näherungsverfahren zur Ermittlung von Isotopieeffekten in natürlichen Systemen gerichtet. In dem Plenarvortrag von *R. Haberlandt* (ZfI) und einem Kurzvortrag von *G. Christoph* (ZfI) wurde gezeigt, daß durch geeignete Modifizierung von Methoden der statistischen Thermodynamik für Isotopenaustauschgleichgewichte das Temperaturregime bei der Bildung der Gesteine angegeben und Aussagen über Ausgangssubstanzen gemacht werden können. Auf die Grenzen und Möglichkeiten der Theorie wurde hingewiesen.

Im ersten seiner beiden Vorträge ging *M. Wolfsberg* (Irvine, USA) auf Anomalien bei Sauerstoffisotopenfraktionierungen ein, die im Ergebnis von Reaktionen bei Funkenentladungen erhalten worden sind. In den USA gelang es, derartige Anomalien, wie sie bisher nur bei Supernovatests betrachtet wurden, im Labormaßstab zu erzeugen. Der zweite Vortrag behandelte die genaue Berechnung von reduzierten Zustandssummenverhältnissen für kleine Moleküle mit Hilfe der Störungstheorie. *J. Heybey* (ZfI) und *H.-L. Vörtler* (ZfI) gaben Zustandsgleichungen für dichte isotopensubstituierte Fluida nach einer von ihnen entwickelten modifizierten Zelltheorie an. Auf Adsorptions- und spezielle Isotopieeffekte unter geologischen Bedingungen gingen *P. Harting, J. Heybey* u. a. (ZfI) bei ihren experimentellen und theoretischen Untersuchungen ein. Mit der theoretisch begründeten Inkrementenmethode, die von *H. Schütze* (ZfI) entwickelt und vorgestellt wurde, ist ein praktikables Näherungsverfahren für die Sauerstoff-Isotopengeothermometrie sowohl an Mineralpaaren als auch bei intramolekularem Isotopenaustausch gegeben.

Von besonderer Bedeutung für die Einordnung geologischer Prozesse sind die in ihrer Interpretation nicht problemlosen Untersuchungen zur Radiogeochronologie und die radiogener Elemente, denen sich eine Reihe von Beiträgen, darunter zwei Plenarvorträge, widmeten. Verstärkt gehen die Bemühungen dahin, durch Kombination verschiedener radiogeochronologischer Methoden unter Einbeziehung von Variationen stabiler Isotope die geologische Aussagemöglichkeit zu verbessern. *J. Šukoljukov* (Moskau, SU) zeigte in seinem Plenarvortrag sehr anschaulich anhand eigener und Literaturdaten zur stofflich-zeitlichen Entwicklung unseres Sonnensystems, daß dieses in seinem frühen Stadium isotopisch nicht homogen war. Diese Inhomogenitäten lassen sich anhand von δD-, $\delta^{15}N$-, $\delta^{18}O$- und δ-Edelgaswerten aus verschiedenen mineralischen Bereichen von Meteoriten ableiten. In einem weiteren Plenarvortrag stellte *B. Amov* (Sofia, Bulgarien) Arbeiten seines Laboratoriums zur Pb/Pb-Radiogeochronologie dar. Obgleich die Pb/Pb-Alter nicht die Genauigkeit der mit der Sm/Nd- und Rb/Sr-Methode ermittelten erreichen, konnte gezeigt werden, daß nach dem von *B. Amov* entwickelten halbempirischen integrativen Modell auch bei Einzelproben Aussagen zur Herkunft des Materials gegeben werden können.

Mit der Rb/Sr-Altersdatierung erzgebirgischer Metamorphite (Phyllite, Glimmerschiefer, Gneise) beschäftigte sich *J. Pilot* (Bergakademie Freiberg) in seinem Vortrag und wies auf die Problematik der Interpretation von Altersangaben bei Metamorphiten hin. Zur Genese der west- und mittelerzgebirgischen Granite anhand von anfänglichen $^{87}Sr/^{86}Sr$-Verhältnissen in Verbindung mit ^{18}O- und Spurenelementgehalten diskutierten *H. Gerstenberger* u. a. (ZfI) für die Jüngeren und Älteren Granite als Quelle ein einheitliches Stammagma. Auch *T. Kaemmel* (Martin-Luther-Universität Halle) macht in einer Systematik von Altersdaten erzgebirgischer und vogtländischer Granite nach unterschiedlichen Methoden verschiedener Laboratorien auf die Schwierigkeit der Interpretation von Altersangaben aufmerksam. Über eine Möglichkeit, die Radioisotopenindikator-Methodik zur Klärung petrologisch-geochemischer Prozesse einzusetzen, informierte *A. Mironov* (Ulan-Ude, SU) am Beispiel des Nuklids ^{195}Au bei Untersuchungen zur Geochemie des Goldes.

Die Vorträge zur Isotopenhydrologie befaßten sich vorwiegend mit der Wechselbeziehung von Niederschlag zu Abfluß, Akkumulation und Luftfeuchte unter Nutzung der Isotopenverhältnisse mehrerer Elemente. Die Bemühungen einer Reihe europäischer Laboratorien, darunter auch das des ZfI, zielen darauf ab, mit ihren langjährigen Untersuchungen zum Niederschlags-Luftfeuchte-Verhalten durch koordiniertes Zusammenwirken zur Klärung der Niederschlagsquellgebiete und des Luftmassentransports in Europa beizutragen. In seinem Plenarvortrag gab *H. Moser* (Neuherberg, BRD) einen Überblick über die kombinierte Anwendung von Deuterium-, Tritium-, ^{13}C-, ^{14}C-, ^{18}O- und ^{85}Kr-Untersuchungen, wie sie am Institut für Radiohydrometrie Neuherberg (BRD) praktiziert wird. Neuere Beispiele von isotopenhydrologischen Untersuchungen zur Akkumulation auf temperierten und antarktischen Gletschern, zur Niederschlags-Abfluß-Beziehung und zu Altersdatierungen von Wasserkörpern deuteten die Vielfalt der Arbeiten dieses Instituts an.

M. Grosvald (Moskau, SU) wertete ^{18}O-Variationen aus Eiskernen der Tiefbohrung Vostok (sowjetische Antarktisstation) und anderer Eistiefbohrungen aus und leitete daraus Aussagen zu paläoklimatischen und glazialen Vorgängen ab. Anhand hydrochemischer und Isotopendaten charakterisierten sowohl *W. Richter* u. a. (ZfI) als auch *R. Krouse* (Calgary, Kanada) die polaren Seen der Schirmacher-Oase (Ost-Antarktika) bzw. stratiforme arktische Seen Nordkanadas. Aus Salinitäts-, Temperatur- und Isotopenmessungen konnten der Chemismus und das thermische Regime dieser polaren Seen erklärt werden. Aus δD- und $\delta^{18}O$-Werten und daraus berechneten Temperaturverhältnissen schlossen *W.-D. Hermichen* u. a. (ZfI) für die Basis des antarktischen Vorgebirgsgletschers südlich der Schirmacher-Oase auf pleistozänes Relikteis, das sich aufgrund des subglazialen Profils erhalten hat. Die Tritiumverteilung im antarktischen Niederschlag untersuchte

D. Hebert (Bergakademie Freiberg) mit einem Mehr-Box-Modell und diskutierte einen separaten Höhen- und Kontinentaleffekt. In Auswertung einer fünfjährigen Niederschlags- und Luftfeuchtesammlung (Gebiet Schirmacher-Oase) diskutierte *P. Kowski* (ZfI) die Deuteriumvariation, wobei auf den Einfluß von Schneeverdunstung und extremem Kondensationsniveau bei der Interpretation hingewiesen wird. *C. Sonntag* u. a. (Heidelberg, BRD) und *K. Rozanski* (Krakow, Polen) bestimmten aus mehrjährigen globalen Meßreihen von Niederschlägen und Luftfeuchte gemäßigter Breiten anhand der D- und ^{18}O-Variationen klimatische Kurz- und Langzeitschwankungen. Die Isotopenvariationen im Wasser lassen sich direkt in organischem Material (z.B. Pflanzen) wiederfinden, so daß daraus paläoklimatische Aussagen abgeleitet werden können. In einigen Vorträgen wurde auf die Anwendung bestimmter Isotope, wie Tritium (*R. Trettin*, ZfI; *H. Jordan* u. a., Bergakademie Freiberg), Deuterium und ^{18}O (*H.-D. Schmiedl* u. a., KFI, Sondershausen), ^{90}Sr und ^{137}Cs (*D. Weiß* u. a., SAAS, Berlin), bei der Lösung hydrogeologischer Probleme eingegangen.

Als methodisch originell darf ein von *G. Müller* u. a. (ZfI) vorgestelltes Verfahren zur Freisetzung von Gasen aus Mineralen mittels eines CO_2-Lasers eingeschätzt werden. Erste Ergebnisse an ^{18}O-Messungen wurden im Vortrag diskutiert. Über Isotopenanalysen an Gasproben im Nanomol-Bereich berichtete *S. Halas* (Lublin, Polen). Die gaschromatographische Trennung von CH_4 und N_2 aus Erdgasen für die Isotopenanalyse erläuterte *G. Hut* (Groningen, Niederlande). *G. Meier* (ZfI) informierte über den Einsatz von Restgas-Quadrupolmassenspektrometern für die Isotopenanalyse leichter Elemente.

Die Posterbeiträge ordneten sich ergänzend in die Thematik der Vorträge ein. Methodisch probenchemische Arbeiten (Gasextraktion an geologischen Proben, Meßgasgewinnung aus Erdgasen, Verfahren zur Bestimmung des Druckeinflusses auf isotopische Systeme, Gasanalysen) und Beiträge, die sich an Vortragsthemen anlehnen (Isotopengeochemie zur Granitgenese und zu Vererzungsvorgängen, Antarktisthematik, Isotopieeffekte in natürlichen Systemen), standen im Vordergrund.

In den wissenschaftlichen Vorträgen und in der Diskussion der Arbeitstagung wurden von den Teilnehmern wertvolle Denkanstöße gegeben und Tendenzen in relevanten Problemen der Isotopenforschung aufgezeigt. Der Veranstalter wird die Vorträge und Poster in den ZfI-Mitteilungen Nr. 84/85 (1984) publizieren.

G. Strauch

Eingegangen am 2. 3. 1984

Isotopenpraxis 20 (1984) 9, S. 361 bis 364

Buchbesprechungen

Stochastische digitale Verarbeitung stetiger Signale

I. J. Bilinski, A. K. Mikelson. Riga: Sinatje 1983. 292 S., 88 Abb., 198 Lit.; geb. Rubel 1,60; M 8,–

Der Inhalt des vorliegenden Buches umfaßt zwei früher häufig getrennt betrachtete Gebiete, die Analog-Digital-Wandlung von Signalen und die digitale Signalverarbeitung. Die Bedeutung und auch Verknüpfung beider Gebiete ist in den letzten Jahren enorm gewachsen. Das ist wesentlich auf das Vordringen der Mikroelektronik und darauf aufbauenden digitalen Verarbeitungssystemen zurückzuführen. Die Autoren vertreten die Ansicht, daß für die effektive Nutzung dieser Mittel die weitgehende Übereinstimmung von Methoden der A/D-Wandlung und Besonderheiten der digitalen Weiterverarbeitung eine wichtige Voraussetzung ist.

Die Spezifik der Darlegungen besteht darin, daß für die Behandlung dieser Probleme stochastische Denkweisen vorgestellt werden. Dabei geht es nicht um die Analyse von mit Rauschen überlagerten Signalen, sondern um die Nutzung stochastischer Methoden für die Realisierung der grundlegenden Operationen der A/D-Wandlung, der Diskretisierung und der Quantifizierung. Die Kapitel behandeln im einzelnen:

1: Stochastische Signalumwandlung, 2: Stochastische Diskretisierung, 3: Stochastische Quantifizierung 1. Art, 4: Stochastische Quantifizierung 2. Art, 5: Anwendung zufälliger Prozesse, 6: Rekonstruktion stochastisch veränderter Signale, 7: Analyse periodischer und zufälliger Signale.

Das Buch ist keine Einführung in die digitale Signalverarbeitung, sondern behandelt eine der möglichen Entwicklungsrichtungen dieser Thematik. Deshalb kann es vor allem dem bereits auf diesem Gebiet Tätigen zum Studium der vorgestellten Denkweise empfohlen werden. Weitere interessierte Leser sollten zum besseren Verständnis zumindest über Kenntnisse der Wahrscheinlichkeitsrechnung und der Theorie stochastischer Prozesse verfügen.

M. Remer

Physiological Plant Ecology III
Responses to the Chemical and Biological Environment

Ed.: O. L. Lange, P. S. Nobel, C. B. Osmond, H. Ziegler. Berlin – Heidelberg – New York: Springer 1983. 799 S., 104 Abb., 70 Tab. (Encyclopedia of Plant Physiology, New Series, Vol. 12, Part C); geb. DM 298,–; US \$ 119,20

Der Teilband III der physiologischen Pflanzenökologie – oder der Ökophysiologie, wie in einigen Überschriften genannt – hat nur die Reaktionen der Pflanzen auf die chemische und biologische Umwelt zum Gegenstand, allerdings mit Ausnahme der „Chemikalien“ Wasser und Kohlendioxid, die ja bereits im Teilband II behandelt wurden. Man erwartet also in diesem Teilband alle für das Wachstum notwendigen Anionen (Nitrate, Sulfate, Phosphate) und Kationen (hauptsächlich die Übergangselemente und Kalium, Kalzium, Aluminium u. a.). Aber die Pflanzen stehen auch im Wechselverhältnis untereinander, insbesondere mit den Mikroorganismen und Viren in der Bodenzone, und mit den samen- und pflanzenfressenden Tieren; dieses Wechselverhältnis sowie die Flechten, fleischfressenden Pflanzen, die N_2-Fixierung und Parasiten werden als biologische Umwelt im Detail vorgestellt.

An dem Teilband haben 24 Autoren in 19 Beiträgen einschließlich einer allgemein gehaltenen Einführung mitgewirkt und eine Darstellung dieses sehr komplexen Themenkreises geschaffen, wie sie wohl noch am besten zu überschauen ist — ein Verdienst der Herausgeber. Die anorganischen Ionen in Gemeinschaft mit Wasser und Kohlendioxid sichern je nach den Konzentrationen das Wachstum und stabilisieren die ionischen, osmotischen und die pH-Wert-Verhältnisse in der Bodenzone. Bei den Übergangselementen werden bei Überschreitung bestimmter Toleranzkonzentrationen rasch Bereiche berührt, die im weitesten Sinne zur Toxikologie zu rechnen sind. Die Überschreitung wird an manchen Stellen sogar bewußt vorgenommen, beispielsweise beim Silber (S. 287) mit seiner Giftwirkung auf Pflanzenwurzeln. Das andeutungsweise gewählte Einteilungsprinzip der Übergangselemente nach dem HSAB-Konzept (hard and soft acids and bases) ist zu undeutlich dargestellt; hier hätte man entweder etwas mehr bringen müssen oder zumindest dem Leser sagen sollen, wo er weitere Einzelheiten darüber erfahren kann (*R. G. Pearson* und seine Arbeiten Mitte der 60er Jahre). Bei der Beschreibung der N-Dynamik in der Bodenzone hätten die in jüngster Zeit von seiten der landwirtschaftlichen Forschung und Praxis inaugurierten $N_{mineral.}$-Dynamikmodelle (beispeilsweise Universität Hannover) erwähnt werden sollen, ganz abgesehen von den herrschenden Gesetzmäßigkeiten der N-Isotopenverhältnisse in der Bodenzone oder bei der N_2-Fixierung (ohne ^{15}N-Markierung). Bemerkenswert ist die Anwendung der r- und K-Strategie auf das Konkurrenzverhalten von Viren (S. 537ff.), der Kampf der Gene in der Virus-Ökologie; hier wurden Bezüge zur Evolution von Bakterien und Viren insgesamt geschaffen oder vielleicht eine Ökologie der Prokaryonten versucht?

Dieser Teilband knüpft in gewohnter Zuverlässigkeit, Aktualität und „gekonnt" abgefaßter Art an seine beiden Vorgänger an: Er liefert viele Informationen, die auf andere Weise nur durch eine sehr kostenaufwendige und auch nervenaufreibende Sucharbeit zusammengetragen werden könnten. Mit insgesamt 3626 Literaturhinweisen (42% aus den Jahren 1976—1983) ist die Aktualität auch quantitativ begründet. Wie jedes Handbuch wird auch dieser Teilband von seinem kaum näher zu definierenden Nutzerkreis zu Rate gezogen werden — Botaniker, Chemiker, Mikrobiologen, Physiologen, Landwirte, Pedologen und nicht zuletzt Umweltforscher werden auf jeden Fall unter ihnen sein.

H. Hübner

Steric Effects in Biomolecules

Hrsg.: G. Náray-Szabó. Budapest: Akademiai Kiadó 1982. 417 S., 165 Abb., 65 Tab. (alle Beiträge in englischer Sprache); kartoniert M 95,—

In Eger (Ungarn) fand vom 5.—8. Oktober 1981 ein Symposium gleichen Titels statt, und das Buch enthält eine Auswahl der Berichte. Sie wurden vom Herausgeber nach den darin enthaltenen Molekeln geordnet: *i*) kleinere Molekeln, *ii*) DNA, RNA, *iii*) Proteine; das sind hier insbesondere Tetracyclin-Antibiotika, Sulfonamide, Catecholaminderivate sowie andere Neurotransmitter und ihre Analoga, Anästhetika, Ionenträger, Neurotoxine, spezielle Hormone und Enzyme. Es wird über theoretische und/oder experimentelle Untersuchungen der Struktur und Konformation sowie der Reaktivität solcher Verbindungen berichtet. Die Autoren sind Spezialisten, die enge Ausschnitte ihrer Arbeit vorstellen. Einige der 27 Aufsätze sind aber auch breiter angelegt. So schreiben *W. G. Richards* und *E. Pretsch* über einige Probleme des Entwerfens von Arzneimitteln (*R.* über die Berücksichtigung von Nicht-Gleichgewichtskonformationen; *P.* über die Entwicklung von Ionenträgern, basierend auf quantentheoretischen Berechnungen), *F. Cramer* diskutiert die Ursachen der niedrigen Fehlerquote enzymatischer Reaktionen am Beispiel der Aminoacylierung von *t*RNA, *H. Wamhoff* gibt einen Überblick zur Photocycloaddition von Nucleinsäuren und deren Analoga, und *R. Zahradnik* untersucht theoretische Probleme der Biokatalyse.

Nur zwei Aufsätze sind dem „sterischen Effekt" als dem eigentlichen Thema des Buches zuzuordnen, nämlich *B. Pullman*: Über die Verfügbarkeit wichtiger Akzeptorplätze verschiedener Regionen der DNA und *L. Ötvös*: Zu sterischen Substituenteneffekten.

Der teilweise widersprüchliche Umgang einzelner Autoren mit dem Begriff des sterischen Effekts läßt vermuten, daß die Beiträge nicht nach strengen Kriterien begutachtet wurden. So liest man beispielsweise auch von „konformationellen Eigenschaften der H_2-Molekel" (S. 3) oder daß die Röntgenkristallstrukturanalyse einen Punkt der (intramolekularen) Potentialfläche liefere (S. 65). Die H_2-Molekel hat aber überhaupt keine konformationellen Freiheitsgrade, und die Röntgenkristallstruktur einer Molekel entspricht nur einem Punkt im Koordinatenraum. (Letzteres gilt für jede Molekelgeometrie.) Intramolekulare Potentiale oder einzelne Punkte derselben werden nicht mittels Röntgenkristallstrukturanalyse bestimmt.

Trotz kleinerer Mängel ist das Buch für alle die Leser von Interesse, die sich mit der Stereochemie biologisch aktiver Substanzen beschäftigen.

P. Scharfenberg

Das Geheimnis des Grünen Planeten
Natur und Mensch —
Der ökologische Frieden und seine Bedrohung

J. Illies. Frankfurt/Main: Umschau-Verlag 1982. 248 S., 45 Farbbilder, 29 schwarz/weiß-Abb.; geb. DM 39,90

Isaac Newton hat mit seiner *lex tertia*, wonach eine Wirkung stets einer Gegenwirkung gleich sei, den Übergang von der Mechanik des einzelnen Massenpunktes zur Mechanik der zusammengesetzten Systeme vollzogen. Zur Beschreibung des wohlabgestimmten Wechselspiels der Enzyme in einem Organismus oder der vielseitigen Wechselbeziehungen von belebter und unbelebter Natur in einem Ökosystem reicht eine so einfache Formulierung nicht aus. Bei solchen komplexen Vorgängen und Zusammenhängen in der Natur können wir nur einige Aspekte herausgreifen, die allerdings bei einer Überhöhung und Überbetonung leicht zu verzerrten Bildern führen und führen müssen.

Professor *Joachim Illies* war Biologe, insbesondere Limnologe; er starb kurz nach Erscheinen dieses Buches. Seine Hauptforderung, ja sein Vermächtnis, ist die Suche nach einem neuen Verhältnis zwischen dem Menschen und der Natur. Für einen Biologen eine naheliegende, in der Gegenwart leicht verständliche und anzuerkennende Forderung, wenn wir uns die vielen Beispiele eines rücksichtslosen Umgangs mit der Natur vergegenwärtigen: Monokulturen in der Landwirtschaft, Mülldeponie-Skandale, Robbenabschlachtungen, saurer Regen usw. . . . eine kaum endende Reihe von Beispielen.

Das Buch ist in 4 große Teile gegliedert: der grüne Planet, die Evolution des Lebendigen, der Mensch und schließlich die Natur und der Mensch, der ökologische Frieden und seine Bedrohung. Jeder dieser Teile wird mit den Augen eines Biologen betrachtet und verführt den Leser dazu, immer wieder das Geheimnisvolle des Lebens, das Einmalige unseres Planeten zu durchleben. Oft sind allerdings auch Elemente eines romantisch-naturphilosophischen Glaubens und einer mystischen Verehrung enthalten, die in eine unverständliche Ablehnung der Evolution — oder besser gesagt — ihrer einfachen Mechanismen mündet. Sätze wie „Leben ist mehr als Chemie" oder„ . . . es gibt eine Stilistik der organischen Gestalten von einer in sich selbst ruhenden, keinem mechanischen Zweck zuzuordnenden Schönheit der Farbe und Muster, die mit dem Geheimnis des Lebendigen direkt verknüpft sind . . ." (S. 71) u. a. lehnen sich stark an *Friedrich W. J. Schelling* an, führen aber kaum zu neuen Einsichten und Zusammenhängen, wie sie doch so vielversprechend angekündigt wurden. Um beispielsweise ein bestimmtes, in sich abgeschlossenes Ökosystem verstehen zu wollen, bedarf es keiner übersinnlichen Betrachtungen, allerdings, wie *John Muir* 1911 schon einmal sagte „When we try to pick out anything by itself, we find it hitched to everything else in the universe".

Einige Ungenauigkeiten haben sich eingeschlichen: Die *Ediacara*-Formation ist rund 150 Millionen Jahre früher nachweisbar als angegeben (S. 69); die Isua-Sedimente sind auf die $^{12}C/^{13}C$-Verhältnisse und nicht $^{12}C/^{14}C$-Verhältnisse hin untersucht worden (S. 67); die sogenannte *Green-Banks*-Gleichung (S. 28) wird mehr nach Art einer Science-fiction-Version interpretiert, es ist aber die *Frank-Drake*-Gleichung gemeint, die zwar bestimmte Unsicherheiten enthält, aber nichts Phantastisches an sich hat; der Physiologe *Kielmeyer* hieß *Karl Friedrich* und nicht *H. H.* (S. 63; S. 247).

Unser mit Leben erfüllter grüner Planet Erde setzt sich aus einem Mosaik von Einzelökosystemen zusammen. Wir kennen kein einfaches Prinzip, das in gleicher Weise wie die *lex tertia* ein schönes Bauwerk oder eine neue Brücke berechenbar macht; denn die belebte Natur ist komplexer, auch schöner und geheimnisvoller, bedarf aber keiner *vis vitalis*. Wir stecken zwar noch in den Anfängen eines vielversprechenden naturwissenschaftlichen Zuganges zum Verständnis dieser besonderen Bewegungsformen der Materie, die wir Leben nennen, aber nichts deutet darauf hin, daß dieser Weg sich als falsch erweisen könnte. Die Sprache des Buches und die ausgewählten Bilder sind schön, so schön, wie sich unser Planet eben darzustellen versteht — wir sollten auch besorgt und rücksichtsvoll mit ihm umgehen.

H. Hübner

The Human Environment — Past, Present and Future

M. Eisenbud. Bethesda, MD: National Council on Radiation Protection and Measurements 1983. 40 S., 6 Abb., 4 Fotos (The Lauriston S. Taylor Lecture Series in Radiation Protection and Measurements, Lecture No. 7)

Im Jahre 1964 wurde der Nationalrat für Strahlenschutz und Messungen (National Council on Radiation Protection and Measurements — NCRP) in den USA gegründet. Er ist eine uneigennützige Körperschaft mit dem Ziel der Verbreitung von Informationen über den Strahlenschutz, er arbeitet Empfehlungen aus und kooperiert mit der Internationalen Kommission für Strahlenschutz (International Commission on Radiological Protection — ICRP). Von 1967 bis 1977 war *Lauriston S. Taylor* der erste Vorsitzende des NCRP, und ihm zu Ehren werden anläßlich der Jahresversammlungen des NCRP Vorträge gehalten — die Lauriston S. Taylor Lectures. Im Jahre 1983 fiel diese Ehre auf Dr. *Merril Eisenbud*, Professor für Umweltmedizin der New York Universität. Sein Vortrag hatte die Umwelt des Menschen in Vergangenheit, Gegenwart und Zukunft zum Gegenstand; die kleine Schrift ist der Text dieses Vortrages.

Die Umwelt des Menschen wird auf drei Grundkomponenten und deren gegenseitigen Ausgleich zurückgeführt: die Gesamtheit der Menschen, die Bereitstellung von Wasser und von Nahrung. Und diese Umwelt ist bedroht: in einer hierarchischen Ordnung angeführt durch einen wahnwitzigen Atomkrieg, gefolgt von einer ständig steigenden Bevölkerungszahl, einer zunehmenden Verknappung von Rohstoff- und Energieressourcen und schließlich von der Umweltbelastung. Diese Umweltbelastungen werden beschrieben, wobei allerdings die radioaktive Strahlenbelastung nur als ein Teil der Belastungen dargestellt wird. So wird das CO_2-Problem mit seinen Folgen auf unser Klima behandelt und u. a. festgestellt, daß im Jahre 1981 die Weltgetreide-Reserven nur 15% eines Jahresbedarfes (entsprechend einer 55 Tage andauernden Ernährungsmöglichkeit) ausgemacht haben; eine drastische, kurzzeitige Klimaveränderung hätte verheerende Folgen für uns alle. Betont wird an vielen Stellen dieser Schrift die Tatsache, daß wir zwar eine Reihe technologischer Lösungsvorschläge und auch geeignete Einrichtungen besitzen, aber vielerorts nicht die dazu notwendigen sozialpolitischen und sozialökonomischen Voraussetzungen, um sie anzuwenden. Ein Leben in Frieden, eine verantwortungsbewußte Bevölkerungsentwicklung und ein maßvoller Umgang mit den Ressourcen unserer Erde sind die Fundamente, auf denen alle Umweltentwicklungsziele stehen müssen.

H. Hübner

Enrichment Techniques for Inorganic Trace Analysis

A. Mizuike. Berlin-Heidelberg-New York: Springer 1983. VIII + 144 S., 41 Abb., 49 Tab., 785 Lit. (Chemical Laboratory Practice, Vol. 19); geb. DM 72,—; US-$ 29·80

Die anorganische Spurenanalyse mit Hilfe optischer, elektrochemischer und nuklearer Analysenverfahren erfordert wegen der an sie gestellten hohen Ansprüche in vielen Fällen eine vorherige Anreicherung des zu bestimmenden Elements. Die vorliegende Monographie gibt eine umfassende praxisbezogene Darstellung der dazu geeigneten Arbeitstechniken.

In den drei einführenden Kapiteln des Buches wird auf die allgemeinen Gesichtspunkte der Anreicherung von Spuren und ausführlich auf die Kontamination der Probe, z.B. durch die Laboratoriumsatmosphäre, die benutzten Reagenzien und Gefäße usw. sowie den Verlust der Spur im Anreicherungsprozeß eingegangen, und es werden Möglichkeiten zur Minimierung dieser hauptsächlichen Fehlerquellen der Spurenanalyse mit vorangegangener Anreicherung dargelegt.

Die danach folgenden acht Kapitel behandeln Anreicherungstechniken, die auf der Verflüchtigung, der Flüssig-Flüssig-Extraktion, der selektiven Auflösung, der Fällung (Mitfällung), der elektrochemischen Abscheidung, der Sorption, dem Ionenaustausch, der Verteilungschromatographie, der Flotation, dem Ausfrieren und dem Zonenschmelzen beruhen, sowie ihre Anwendungen. In jedem Kapitel wird der Benutzer zunächst in Abhängigkeit von der Bedeutung der Methode mehr oder weniger ausführlich mit deren Grundlagen vertraut gemacht. Der besondere Wert des Buches liegt vor allem darin, daß im Anschluß daran jeweils eine Vielzahl in der Literatur erschienener Arbeiten, auch in Form übersichtlicher Tabellen, zitiert wird, denen für die unterschiedlichsten Kontaminationen von

Spur- und Matrixelementen detaillierte Angaben zur Anreicherung der Spur und zum benutzten Analysenverfahren zu entnehmen sind. Zwei Anwendungsbeispiele, in denen Anreicherungstechniken für die Spurenanalyse von Wasser- und Gasproben empfohlen werden, sowie ein Anhang u. a. mit nützlichen Tabellen und graphischen Darstellungen zum Ionenaustausch runden den behandelten Stoff ab.

Das Buch ist für die Laboratoriumspraxis eine wertvolle Hilfe, obwohl der Autor auf die Wiedergabe von Arbeitsvorschriften weitestgehend verzichtet. Dadurch kommt das Buch mit einem geringen Umfang aus, was der übersichtlichen, geschlossenen Darstellung der Problematik förderlich ist. Der mit der Spurenanalyse befaßte Chemiker oder Physiker wird durch die Besprechung von nahezu 800 publizierten Arbeiten für die meisten der vor ihm stehenden Aufgaben einen oder mehrere Lösungswege angegeben finden. Die bis 1982 recherchierte Literatur entspricht dem aktuellen Stand: nur 36% der zitierten Arbeiten sind vor 1970 erschienen.

Das Buch spricht einen vielfältigen Interessentenkreis an. Es kann z.B. dem im Umweltschutz tätigen Chemiker zur anorganischen Spurenanalyse von geologischen, biologischen oder hydrologischen Materialien ebenso empfohlen werden wie dem Radiochemiker, der insbesondere zur Anwendung der Extraktion, der Mitfällung und des Ionenaustausches viele Anregungen finden wird. *P. Hecht*

Geological Evolution of the Earth during the Precambrium

L. J. Salop. Übersetz. aus dem Russ. ins Engl. von V. P. Grudina. Berlin-Heidelberg-New York: Springer Verlag 1983. XII + 459 S., 78 Abb., 5 Tab., geb. DM 148,–; US $ 59.20

Man stelle sich eine Enzyklopädie vor, die aus 20 Bänden besteht, und jeder einzelne Band hätte 1000 Seiten. Die Erdgeschichte würde dann wie folgt beschrieben werden können: das Phanerozoikum, also die Zeit der erkennbaren Lebewesen, würde ungefähr im 2. Band auf Seite 350 zu finden sein, das Aussterben der Saurier wäre ungefähr auf Seite 276 des 1. Bandes zu suchen, und die uns überlieferte Menschheitsgeschichte wären nur einzelne Zeilen und Worte auf der ersten Seite des riesigen Werkes, denn eine Seite entspräche ja rund einer viertel Million Jahre. Und der Rest? Der Rest aller übrigen Bände wäre die Geschichte des Präkambriums, 6/7 der Erdgeschichte darstellend – eine unendlich lang erscheinende Zeitspanne.

Die geologische Entwicklung des Präkambriums wird in 6 Kapiteln dargestellt, jedes Kapitel konsequent zweigeteilt in die vorkommenden Gesteine mit ihren stratigraphischen Einheiten und den entsprechenden geologischen Interpretationen, dem eigentlichen Anliegen dieser Monographie. Flankiert werden die 6 Kapitel von einer einführenden Darstellung der grundsätzlichen Probleme und Schwierigkeiten, denen man in der Erforschung des Präkambriums begegnet, und einer abschließenden Betrachtung über Zyklizitäten der geologischen Prozesse und die Periodizitäten der Orogenesen.

Es ist bemerkenswert, daß jede zeitliche Zuordnung in diesem langen Abschnitt der geologischen Entwicklung unserer Erde mit Hilfe der kernphysikalischen Datierungsmethoden vorgenommen wird – das gesamte Instrumentarium dieser Technik muß aufgeboten werden, um eine Aufhellung in der Stratigraphie zu erreichen. Weniger vollständig und kritisch werden die anderen Isotopenmethoden einbezogen, $\delta^{13}C$-Werte der frühesten Lebensspuren, $\delta^{18}O$-Werte der Karbonate, $\delta^{34}S$-Werte der Sulfide und Sulfate aus diesem Zeitraum. Die Arbeiten von *M. Schidlowski* beispielsweise aus den Jahren 1979–1981 zu diesem Gebiet der archaischen Kohlenstoffisotopen-Rekorde sind nicht mit erfaßt, wie überhaupt der geochemische – oder sagen wir besser – biogeochemische Aspekt etwas hintansteht. Die Geschichte unserer Erde kann nur komplex und dynamisch verstanden werden – die Geologie im Wechselverhältnis mit dem Klima, der Atmosphäre und der Biosphäre, manchmal in den Eiszeiten auch mit der Kryosphäre.

Es bleibt zu fragen, was uns bewegt, uns mit diesem archaischen Zeitraum zu beschäftigen. Es sind die ungeheuer reichen Vorkommen an Eisen- und Uranerzen, die aus dieser Zeit stammen, und deren Entstehungsgeschichte man kennen will; aber es sind auch Fragen der reinen Grundlagenforschung, welche Entwicklungsstufen die Atmosphäre in ihrer Zusammensetzung durchgemacht hat. Die Entstehung des Lebens fällt in diese Zeit, eine der aufregendsten Fragen, wie sie gegenwärtig gestellt wird und erste Schritte zu ihrem Verständnis erlaubt.

Die Literatur ist aktuell erfaßt (35,5% aus den Jahren 1976–1983 von insgesamt 645 Zitaten), aber der Strom der Publikationen geht unaufhaltsam weiter. Nur so kann Seite für Seite der vielbändigen Geschichte des Präkambriums mit unseren modernen Methoden untersucht, eingeordnet und verstanden werden.

Der sowjetische Autor *L. J. Salop* hat selbst an vielen Seiten mitgeschrieben und eine Bilanz gezogen, die in diesem Buch vorliegt. Es wird viel beachtet und oft erwähnt werden. *H. Hübner*

Systems Analysis – Modelling – Simulation

Journal of Mathematical Modelling and Simulation in Systems Analysis

Edited at the Central institute of Cybernetics and Information Processes of the Academy of Sciences of the G.D.R. in cooperation with the Mathematical Society of the G.D.R. by Achim Sydow (Berlin) with the assistance of an international team of scientists

Published yearly, 1 volume (4 numbers).
Volume 1, 1984;
16,7 cm by 24,0 cm, Annual list of contents.
Annual subscription M 120,—; single number M 30,—.
Order number: 1097 ISSN 0232-9298

The journal is devoted to art and techniques and applications of modelling and simulation in systems analysis. Systems analysis is an interdisciplinary activity. Articles will be published on systems analysis and systems theory, modelling and mathematical models and tools for decision and control, computer simulation systems, modelling and simulation in environmental protection, macroeconomics, agricultural production, regional planning, biosciences, traffic control, large engineering systems, resources papers, distribution, management systems etc. In addition to original survey articles and reports on conferences etc. are published. Finally, books covering the above mentioned and related subjects are reviewed.
Of interest to: Scientists of systems sciences, informatics, control engineering, applied mathematics, operation research, biology, economy, physics, environmental protection agriculture etc.

Orders can be sent

- in the GDR: to a book-shop or to the Akademie-Verlag, DDR-1086 Berlin, Leipziger Str. 3–4, PF-Nr. 1233;
- in the other socialist countries: to a book-shop for foreign language literature or to the competent news-distributing agency;
- in the FRG and Berlin (West): to the wholesale distributing agency Kunst und Wissen, Erich Bieber OHG, Wilhelmstr. 4–6, D-7000 Stuttgart 1;
- in the other Western European Countries: to Kunst und Wissen, Erich Bieber GmbH, Dufourstr. 51, CH-8008 Zürich;
- in other countries: to the Akademie-Verlag, DDR-1086 Berlin, Leipziger Str. 3–4, PF-Nr. 1233.

AKADEMIE-VERLAG · BERLIN
1086-Berlin, Leipziger Straße 3–4

ISOTOPENPRAXIS

Volume 20 — 1984 — Heft 9

Contents

Inhalt

Isotopenpraxis is indexed in Current Contents/PC & ES, Science Citation Index, and INIS Atomindex

Zeitfracht Medien GmbH
Ferdinand-Jühlke-Straße 7
99095 Erfurt, Deutschland
produktsicherheit@kolibri360.de